Modern Physics Simulations
The Consortium for
Upper-Level Physics Software

Douglas Brandt
Department of Physics, Eastern Illinois University
Charleston, Illinois

John R. Hiller
Department of Physics, University of Minnesota
Duluth, Minnesota

Michael J. Moloney
Department of Physics, Rose-Hulman Institute
Terre Haute, Indiana

Series Editors

Robert Ehrlich

William MacDonald

Maria Dworzecka

JOHN WILEY & SONS, INC.
NEW YORK · CHICHESTER · BRISBANE · TORONTO · SINGAPORE

ACQUISITIONS EDITOR Cliff Mills
MARKETING MANAGER Catherine Faduska
PRODUCTION EDITOR Sandra Russell
MANUFACTURING MANAGER Mark Cirillo

This book was set in 10/12 Times Roman by Beacon Graphics and
printed and bound by Hamilton Printing Co. The cover was printed by Phoenix Color.

Recognizing the importance of preserving what has been written, it is a
policy of John Wiley & Sons, Inc. to have books of enduring value published
in the United States printed on acid-free paper, and we exert our best
efforts to that end.

The paper on this book was manufactured by a mill whose forest management programs include
sustained yield harvesting of its timberlands. Sustained yield harvesting principles ensure that
the number of trees cut each year does not exceed the amount of new growth.

Library of Congress Cataloging in Publication Data:
Brandt, Douglas.
    Modern physics simulations: the Consortium for Upper Level Physics Software / Douglas Brandt, John R.
    Hiller, Michael J. Moloney
        p.   cm.
    Includes bibliographical references (p.).
    ISBN 0-471-54882-0 (pbk./disk: alk. paper)
    1. Physics--Computer simulation. 2. Physics--Software.
    I. Hiller, John R.   II. Moloney, Michael.   III. Consortium for Upper Level Physics Software.   IV. Title.
QC52.B73   1995   95-35661
530'.01'13--dc20   CIP

Printed in the United States of America

10   9   8   7   6   5   4   3   2   1

Contents

List of Figures

1

Introduction

"It is nice to know that the computer understands the problem. But I would like to understand it too."

—Eugene P. Wigner, quoted in *Physics Today,* July 1993

1.1   *Using the Book and Software*

The simulations in this book aim to exploit the capabilities of personal computers and provide instructors and students with valuable new opportunities to teach and learn physics, and help develop that all-important, if somewhat elusive, physical intuition. This book and the accompanying diskettes are intended to be used as supplementary materials for a junior- or senior-level course. Although you may find that you can run the programs without reading the text, the book is helpful for understanding the underlying physics, and provides numerous suggestions on ways to use the programs. *If you want a quick guided tour through the programs, consult the "Walk Throughs" in Appendix A.* The individual chapters and computer programs cover mainstream topics found in most textbooks. However, because the book is intended to be a supplementary text, no attempt has been made to cover all the topics one might encounter in a primary text.

Because of the book's organization, students or instructors may wish to deal with different chapters as they come up in the course, rather than reading the chapters in the order presented. One price of making the chapters semi-independent of one another is that they may not be entirely consistent in notation or tightly cross-referenced. Use of the book may vary according to the taste of the student or instructor. Students may use this material as the basis of a self-study course. Some instructors may make homework assignments from the large number of exercises in each chapter or to use them as the basis of student projects. Other instructors may use the computer programs primarily for in-class demonstrations. In this latter case, you may find that the programs are suitable for a range of courses from the introductory to the graduate level.

Use of the book and software may also vary with the degree of computer programming performed by users. For those without programming experience, all the computer simulations have been supplied in executable form, permitting them to be used as is. On the other hand, Pascal source code for the programs has also been provided, and a number of exercises suggest specific ways the programs can be modified. Possible modifications range from altering a single procedure especially set up for this purpose by the author, to larger modifications following given examples, to extensive additions for ambitious projects. However, the intent of the authors is that the simulations will help the student to develop intuition and a deeper understanding of the physics, rather than to develop computational skills.

We use the term "simulations" to refer to the computer programs described in the book. This term is meant to imply that programs include complex, often realistic, calculations of models of various physical systems, and the output is usually presented in the form of graphical (often animated) displays. Many of the simulations can produce numerical output—sometimes in the form of output files that could be analyzed by other programs. The user generally may vary many parameters of the system, and interact with it in other ways, so as to study its behavior in real time. The use of the term simulation should not convey the idea that the programs are bypassing the necessary physics calculations, and simply producing images that look more or less like the real thing.

The programs accompanying this book can be used in a way that complements, rather than displaces, the analytical work in the course. It is our belief that, in general, computational and analytical approaches to physics can be mutually reinforcing. It may require considerable analytical work, for example, to modify the programs, or really to understand the results of a simulation. In fact, one important use of the simulations is to suggest conjectures that may then be verified, modified, or proven false analytically. A complete list of programs is given in Section 1.7.

1.2   Required Hardware and Installation of Programs

The programs described in this book have been written in the Pascal language for MS-DOS platforms. The language is Borland/Turbo Pascal, and the minimum hardware configuration is an IBM-compatible 386-level machine preferably with math coprocessor, mouse, and VGA color monitor. In order to accommodate a wide range of machine speeds, most programs that use animation include the capability to slow down or speed up the program. To install the programs, place disk number 1 in a floppy drive. Change to that drive, and type Install. You need only type in the file name to execute the program. Alternatively, you could type the name of the driver program (the same name as the directory in which the programs reside), and select programs from a menu. A number of programs write to temporary files, so you should check to see if your autoexec.bat file has a line that sets a temporary directory, such as SET TEMP = C:\TEMP. (If you have installed WINDOWS on your PC, you will find that such a command has already been written into your autoexec.bat file.) If no such line is there, you should add one.

Compilation of Programs

If you need to compile the programs, it would be preferable to do so using the
Borland 7.0 (or later) compiler. If you use an earlier Turbo compiler you may run
out of memory when compiling. If that happens, try compiling after turning off
memory resident programs. If your machine has one, be sure to compile with the
math-coprocessor turned on (no emulation). Finally, if you recompile programs
using any compiler other than Borland 7.0, you will get the message: "EGA/VGA
Invalid Driver File" when you try to execute them, because the driver file supplied
was produced using this version of the compiler. In this case, search for the file
BGILINK.pas included as part of the compiler to find information on how to create the EGAVGA.obj driver file. *If any other instructions are needed for installa-
tion, compilation, or running of the programs, they will be given in a README
file on the diskettes.*

1.3 User Interface

To start a program, simply type the name of the individual or driver program,
and an opening screen will appear. All the programs in this book have a common
user interface. Both keyboard and mouse interactions with the computer are possible. Here are some conventions common to all the programs.

Menus: If using the *keyboard*, press **F10** to highlight one of menu boxes, then
use the **arrow** keys, **Home**, and **End** to move around. When you press **Return** a submenu will pull down from the currently highlighted menu option. Use the same keys to move around in the submenu, and press **Return**
to choose the highlighted submenu entry. Press **Esc** if you want to leave the
menu without making any choices.

If using the *mouse* to access the top menu, click on the menu bar to pull
down a submenu, and then on the option you want to choose. Click anywhere
outside the menus if you want to leave them without making any choice.
Throughout this book, the process of choosing submenu entry **Sub** under
main menu entry **Main** is referred to by the phrase "choose **Main | Sub**."
The detailed structure of the menu will vary from program to program, but
all will contain **File** as the first (left-most) entry, and under **File** you will
find **About CUPS**, **About Program**, **Configuration**, and **Exit Program**. The first two items when activated by mouse or arrows keys will
produce information screens. Selecting **Exit Program** will cause the program to terminate, and choosing **Configuration** will present you with a
list of choices (described later), concerning the mode of running the program. In addition to these four items under the **File** menu, some programs
may have additional items, such as **Open**, used to open a file for input, and
Save, used to save an output file. If **Open** is present and is chosen, you
will be presented with a scrollable list of files in the current directory from
which to choose.

Hot Keys: Hot keys, usually listed on a bar at the bottom of the screen, can be
activated by pressing the indicated key or by clicking on the hot key bar
with the mouse. The hot key **F1** is reserved for help, the hot key **F10** activates the menu bar. Other hot keys may be available.

Sliders (scroll-bars): If using the *keyboard*, press **arrow** keys for slow scrolling of the slider, **PgUp/PgDn** for fast scrolling, and **End/Home** for moving from one end to another. If you have more then one slider on the screen then only the slider with marked "thumb" (sliding part) will respond to the above keys. You can toggle the mark between your sliders by pressing the **Tab** key.

If using the *mouse* to adjust a slider, click on the thumb of the slider, drag it to desired value, and release. Click on the arrow on either end of the slider for slow scrolling, or in the area on either side of thumb for fast scrolling in this direction. Also, you can click on the box where the value of the slider is displayed, and simply type in the desired number.

Input Screens: All input screens have a set of "default" values entered for all parameters, so that you can, if you wish, run the program by using these original values. Input screens may include circular radio buttons and square check boxes, both of which can take on Boolean, i.e., "on" or "off," values. Normally, check boxes are used when only one can be chosen, and radio buttons when any number can be chosen.

If using the *keyboard*, press **Return** to accept the screen, or **Esc** to cancel it and lose the changes you may have made. To make changes on the input screen by keyboard, use **arrow** keys, **PgUp**, **PgDn**, **End**, **Home**, **Tab**, and **Shift-Tab** to choose the field you want to change, and use the backspace or delete keys to delete numbers. For Boolean fields, i.e., those that may assume one of two values, use any key except those listed above to change its value to the opposite value.

If you use the *mouse*, click [OK] to accept the screen or [Cancel] to cancel the screen and lose the changes. Use the mouse to choose the field you want to change. Clicking on the Boolean field automatically changes its value to the opposite value.

Parser: Many programs allow the user to enter expressions of one or more variables that are evaluated by the program. The function parser can recognize the following functions: absolute value (abs), exponential (exp), integer or fractional part of a real number (int or frac), real or imaginary part of a complex number (re or im), square or square root of a number (sqr or sqrt), logarithms—base 10 or e (log or ln)—unit step function (h), and the sign of a real number (sgn). It can also recognize the following trigonometric functions: sin, cos, tan, cot, sec, csc, and the inverse functions: arcsin, arccos, arctan, as well as the hyperbolic functions denoted by adding an "h" at the end of all the preceding functions. In addition, the parser can recognize the constants $pi, e, i(\sqrt{-1})$, and rand (a random number between 0 and 1). The operations $+$, $-$, $*$, $/$, $\hat{}$ (exponentiation), and !(factorial) can all be used, and the variables r and c are interpreted as $r = \sqrt{x^2 + y^2}$ and $c = x + iy$. Expressions involving these functions, variables, and constants can be nested to an arbitrary level using parentheses and brackets. For example, suppose you entered the following expression: **h(abs(sin(10*pi* x))−0.5).** The parser would interpret this function as $h(|sin(10\pi x)| - 0.5)$. If the program evaluates this function for a range of *x*-values, the result, in this case, would be a series of square pulses of width 1/15, and center-to-center separation 1/10.

Help: Most programs have context-sensitive help available by pressing the **F1** hot key (or clicking the mouse in the **F1** hot key bar). In some programs help is available by choosing appropriate items on the menu, and in still other programs tutorials on various aspects of the program are available.

1.4   The CUPS Project and CUPS Utilities

The authors of this book have developed their programs and text as part of the Consortium for Upper-Level Physics Software (CUPS). Under the direction of the three editors of this book, CUPS is developing computer simulations and associated texts for nine junior- or senior-level courses, which comprise most of the undergraduate physics major curriculum during those two years. A list of the nine CUPS courses, and the authors associated with each course, follows this section. This international group of 27 physicists includes individuals with extensive backgrounds in research, teaching, and development of instructional software.

The fact that each chapter of the book has been written by a different author means that the chapters will reflect that individual's style and philosophy. Every attempt has been made by the editors to enhance the similarity of chapters, and to provide a similar user interface in each of the associated computer simulations. Consequently, you will find that the programs described in this and other CUPS books have a common look and feel. This degree of similarity was made possible by producing the software in a large group that shared a common philosophy and commitment to excellence.

Another crucial factor in developing a degree of similarity between all CUPS programs is the use of a common set of utilities. These CUPS utilities were written by Jaroslaw Tuszynski and William MacDonald, the former having responsibility for the graphics units, and the latter for the numerical procedures and functions. The numerical algorithms are of high quality and precision, as required for reliable results. CUPS utilities were originally based on the M.U.P.P.E.T. utilities of Jack Wilson and E.F. Redish, which provided a framework for a much expanded and enhanced mathematical and graphics library. The CUPS utilities (whose source code is included with the simulations with this book), include additional object-oriented programs for a complete graphical user interface, including pull-down menus, sliders, buttons, hot-keys, and mouse clicking and dragging. They also include routines for creating contour, two-dimensional (2-D) and 3-D plots, and a function parser. The CUPS utilities have been provided in source code form to enable users to run the simulations under future generations of Borland/Turbo Pascal. If you do run under future generations of Turbo or Borland Pascal on the PC, the utilities and programs will need to be recompiled. You will also need to create a new egavga.obj file which gets combined with the programs when an executable version is created—thereby avoiding the need to have separate (egavga.bgi) driver files. These CUPS utilities are also available to users who wish to use them for their own projects.

One element not included in the utilities is a procedure for creating hard copy based on screen images. When hard copy is desired, those PC users with the appropriate graphics driver (graphics.com), may be able to produce high-quality screen images by depressing the **PrintScreen** key. If you do not have the graphics software installed to get screen dumps, select **Configuration ∣ Print Screen**, and follow the directions. Moreover, public domain software also exists for capturing screen images, and for producing PostScript files, but the user should be aware that such files are often quite large, sometimes over 1 MB, and they require a PostScript printer driver to produce.

One feature of the CUPS utilities that can improve the quality of hard copy produced from screen captures is a procedure for switching colors. This capability is important because the grey scale rendering of colors on black-and-white printers may create poor contrasts if the original (default) color assignments are used. To access the CUPS utility for changing colors, the user need only choose **Configuration** under the **File** menu when the program is first initiated, or at any later time. Once you have chosen **Configuration**, to change colors you need to click the mouse on the **Change Colors** bar, and you will be presented with a 16 by 16 matrix of radio buttons that will allow you to change any color to any other color, or else to use predefined color switches, such as a color "reversal," or a conversion of all light colors to black, and all dark colors to white. (The screen captures given in this book were produced using the "reverse" color map.) Any such color changes must be redone when the program is restarted.

Other system parameters may likewise be set from the **File | Configuration** menu item. These include the path for temporary files that the program may create (or want to read), the mouse "double click" speed—important for those with slow reflexes—an added time delay to slow down programs on computers that are too fast, and a "check memory" option—primarily of interest to those making program modifications.

Those users wishing more information on the CUPS utilities should consult the CUPS Utilities Manual, written by Jaroslaw Tuszynski and William MacDonald, published by John Wiley and Sons. However, it is not necessary for casual users of CUPS programs to become familiar with the utilities. Such familiarity would only be important to someone wishing to write their own simulations using the utilities. The utilities are freely available for this purpose, for unrestricted noncommercial production and distribution of programs. However, users of the utilities who wish to write programs for commercial distribution should contact John Wiley and Sons.

1.5   Communicating With the Authors

Users of these programs should not expect that run-time errors will never occur! In most cases, such run-time errors may require only that the user restart the program; but in other cases, it may be necessary to reboot the computer, or even turn it off and on. The causes of such run-time errors are highly varied. In some cases, the program may be telling you something important about the physics or the numerical method. For example, you may be trying to use a numerical method beyond its range of applicability. Other types of run-time errors may have to do with memory or other limitations of your computer. Finally, although the programs in this book have been extensively tested, we cannot rule out the possibility that they may contain errors. (Please let us know if you find any! It would be most helpful if such problems were communicated by electronic mail, and with complete specificity as to the circumstances under which they arise.)

It would be best if you communicated such problems directly to the author of each program, and simultaneously to the editors of this book (the CUPS Directors), via electronic mail—see addresses listed below. Please feel free to communicate any suggestions about the programs and text which may lead to improvements in future editions. Since the programs have been provided in source code form, it will be possible for you to make corrections of any errors that you or we find in the future—provided that you send in the registration card at the back of the book, so that you can be notified. The fact that you have the source code will also allow you to make modifications and extensions of the programs. We can assume no responsibility for errors that arise in programs that you have modified. In fact, we strongly urge you to change the program name, and to add a documentary note at the beginning of the code of any modified programs that alerts other potential users of any such changes.

1.6   CUPS Courses and Developers

- **CUPS Directors**
  Maria Dworzecka, George Mason University (cups@gmu.edu)
  Robert Ehrlich, George Mason University (cups@gmu.edu)
  William MacDonald, University of Maryland (w_macdonald@umail.umd.edu)

- **Astrophysics**
  J. M. Anthony Danby, North Carolina State University (n38hs901@ncuvm.ncsu.edu)
  Richard Kouzes, Battelle Pacific Northwest Laboratory (rt_kouzes@pnl.gov)
  Charles Whitney, Harvard University (whitney@cfa.harvard.edu)

- **Classical Mechanics**
  Bruce Hawkins, Smith College (bhawkins@smith.bitnet)
  Randall Jones, Loyola College (rsj@loyvax.bitnet)

- **Electricity and Magnetism**
  Robert Ehrlich, George Mason University (rehrlich@gmuvax.gmu.edu)
  Lyle Roelofs, Haverford College (lroelofs@haverford.edu)
  Ronald Stoner, Bowling Green University (stoner@andy.bgsu.edu)
  Jaroslaw Tuszynski, George Mason University (cups@gmuvax.gmu.edu)

- **Modern Physics**
  Douglas Brandt, Eastern Illinois University (cfdeb@ux1.cts.eiu.edu)
  John Hiller, University of Minnesota, Duluth (jhiller@d.umn.edu)
  Michael Moloney, Rose Hulman Institute (moloney@nextwork.rose-hulman.edu)

- **Nuclear and Particle Physics**
  Roberta Bigelow, Willamette University (rbigelow@willamette.edu)
  John Philpott, Florida State University (philpott@fsunuc.physics.fsu.edu)
  Joseph Rothberg, University of Washington (rothberg@phys.washington.edu)

- **Quantum Mechanics**
  John Hiller, University of Minnesota Duluth (jhiller@d.umn.edu)
  Ian Johnston, University of Sydney (idj@suphys.physics.su.oz.au)
  Daniel Styer, Oberlin College (dstyer@physics.oberlin.edu)

- **Solid State Physics**
  Graham Keeler, University of Salford (g.j.keeler@physics.salford.ac.uk)
  Roger Rollins, Ohio University (rollins@chaos.phy.ohiou.edu)
  Steven Spicklemire, University of Indianapolis (steves@truevision.com)

- **Thermal and Statistical Physics**
  Harvey Gould, Clark University (hgould@vax.clarku.edu)
  Lynna Spornick, Johns Hopkins University
  Jan Tobochnik, Kalamazoo College (jant@kzoo.edu)

- **Waves and Optics**
  G. Andrew Antonelli, Wolfgang Christian, and Susan Fischer, Davidson College (wc@phyhost.davidson.edu)
  Robin Giles, Brandon University (giles@brandonu.ca)
  Brian James, Salford University (b.w.james@physics.salford.ac.uk)

1.7   *Descriptions of all CUPS Programs*

Each of the computer simulations in this book (as well as those in the eight other books comprised by the CUPS Project) are described below. The individual headings under which programs appear correspond to the nine CUPS courses. In several cases, programs are listed under more than one course. The number of programs listed under the Astrophysics, Modern Physics, and Thermal Physics courses is appreciably greater than the others, because several authors have opted to subdivide their programs into many smaller programs. Detailed inquiries regarding CUPS programs should be sent to the program authors.

ASTROPHYSICS PROGRAMS

STELLAR (Stellar Models), written by Richard Kouzes, is a simulation of the structure of a static star in hydrodynamic equilibrium. This provides a model of a zero age main sequence star, and helps the user understand the physical processes that exist in stars, including how density, temperature, and luminosity depend on mass. Stars are self-gravitating masses of hot gas supported by thermodynamic processes fueled by nuclear fusion at their core. The model integrates the four differential equations governing the physics of the star to reach an equilibrium condition which depends only on the star's mass and composition.

EVOLVE (Stellar Evolution), written by Richard Kouzes, builds on the physics of a static star, and considers (1) how a gas cloud collapses to become a main sequence star, and (2) how a star evolves from the main sequence to its final demise. The model is based on the same physics as the STELLAR program. Starting from a diffuse cloud of gas, a protostar forms as the cloud collapses and reaches a sufficient density for fusion to begin. Once a star reaches equilibrium, it remains for most of its life on the main sequence, evolving off after it has consumed its fuel. The final stages of the star's life are marked by rapid and dramatic evolution.

BINARIES is the driver program for all Binaries Programs (**VISUAL1, VISUAL2, ECLIPSE, SPECTRO, TIDAL, ROCHE, and ACCRDISK**).

VISUAL1 (Visual Binaries—Proper Motion), written by Anthony Danby, enables you to visualize the proper motion in the sky of the members of a visual binary system. You can enter the elements of the system and the mass ratio, as well as the speed at which the center of mass moves across the screen. The program also includes an animated three-dimensional demonstration of the elements.

VISUAL2 (Visual Binaries—True Orbit), written by Anthony Danby, enables you to select an apparent orbit for the secondary star with arbitrary eccentricity, with the primary at any interior point. The elements of the orbit are displayed. You can see the orbit animated in three dimensions, or can make up a set of "observations" based on the apparent orbit.

ECLIPSE (Eclipsing Binaries), written by Anthony Danby, shows simultaneously either the light curve and the orbital motion or the light curve and an animation of the eclipses. You can select the elements of the orbit and radii and magnitudes of the stars. A form of limb-darkening is also included as an option.

SPECTRO (Spectroscopic Binaries), written by Anthony Danby, allows you to select the orbital elements of a spectroscopic binary, and then shows simultaneously the velocity curve, the orbital motion, and a moving spectral line.

TIDAL (Tidal Distortion of a Binary), written by Anthony Danby, models the motion of a spherical secondary star around a primary that is tidally distorted by the secondary. You can select orbital elements, masses of the stars, a parameter describing the tidal lag, and the initial rate of rotation of the primary. The equations are integrated over a time interval that you specify. Then you can see the changes of the orbital elements, and the rotation of the primary, with time. You can follow the motion in detail around each revolution, or in a form where the equations have been averaged around each revolution.

ROCHE (The Photo-Gravitational Restricted Problem of Three Bodies), written by Anthony Danby, follows the two-dimensional motion of a particle that is subject to the gravitational attraction of two bodies in mutual circular orbits, and also, optionally, radiation pressure from these bodies. It is intended, in part, as background for the interpretation of the formation of accretion disks. Curves of zero velocity (that limit regions of possible motion) can be seen. The orbits can also be followed using Poincaré maps.

ACCRDISK (Formation of an Accretion Disk), written by Anthony Danby, follows some of the dynamical steps in this process. The dynamics is valid up to the initial formation of a hot spot, and qualititative afterward.

NBMENU is the driver program for all programs on the motion of N interacting bodies: **TWO-GALAX, ASTROIDS, N-BODIES, PLANETS, PLAYBACK, and ELEMENTS.**

TWOGALAX (The Model of Wright and Toomres), written by Anthony Danby, is concerned with the interaction of two galaxies. Each consists of a central gravitationally attracting point, surrounded by rings of stars (which are attracted, but do not attract). Elements of the orbits of one galaxy relative to the other are selected, as is the initial distribution and population of the rings. The motion can be viewed as projected into the plane of the orbit of the galaxies, or simultaneously in that plane and perpendicular to it. The positions can be stored in a file for later viewing.

ASTROIDS (N-Body Application to the Asteroids), written by Anthony Danby, uses the same basic model, but a planet and a star take the place of the galaxies and the asteroids replace the stars. Emphasis is on asteroids all having the same period, with interest on periods having commensurability with the period of the planet. The orbital motion of the system can be followed. The positions can be stored in a file for later viewing. An asteroid can be selected, and the variation of its orbital elements can then be followed.

NBODIES (The Motion of N Attracting Bodies), written by Anthony Danby, allows you to choose the number of bodies (up to 20) and the total energy of the system. Initial conditions are chosen at random, consistent with this energy, and the resulting motion can be observed. During the motion various quantities, such as the kinetic energy, are displayed. The positions can be stored in a file for later viewing.

PLANETS (Make Your Own Solar System), written by Anthony Danby, is similar to the preceding program, but with the bodies interpreted as a star with planets. Initial conditions are specified through the choice of the initial elements of the planets. The positions can be stored in a file for later viewing.

PLAYBACK, written by Anthony Danby, enables a file stored by one of the preceding programs to be viewed.

ELEMENTS (Orbital Elements of a Planet), written by Anthony Danby, shows a three-dimensional animation that can be viewed from any angle.

GALAXIES is the driver program for Galactic Kinematics Programs: **ROTATION, OORTCONS, and ARMS21CM**.

ROTATION (The Rotation Curve of a Galaxy), written by Anthony Danby, first prompts you to "design" a galaxy, consisting of a central mass and up to five spheroids (that can be visible or invisible). It then displays the galaxy and can show the animated rotation or the rotation curve.

OORTCONS (Galactic Kinematics and Oort's Constants), written by Anthony Danby, allows you to design your galaxy, choose the location of the "sun" and a local region around it, and the to observe the kinematics in this region. It also shows graphs of radial velocity and proper motion in comparison with the linear approximation, and computes the Oort constants.

ARMS21CM (The Spiral Structure of a Galaxy), written by Anthony Danby, allows you to design your galaxy, construct a set of spiral arms, and select the position of the "sun." Then, for different galactic longitudes, you can see observed profiles of 21 cm lines.

ATMOS (Stellar Atmospheres), written by Charles Whitney, permits the user to select a constellation, see it mapped on the computer screen, point to a star, and see it plotted on a brightness-color diagram. The user's task is to build a model atmosphere that imitates the photometric properties of observed stars. This is done by specifying numerical values for three basic stellar parameters: radius, mass, and luminosity. The program then builds the model and displays it on the brightness-color diagram, and it also plots the spectrum and the detailed thermodynamic structure of the atmosphere. With this program the user may investigate the relation between stellar parameters and the thermal properties of the gas in the atmosphere. Two atmospheres may be superposed on the graphs, for easier comparison.

PULSE (Stellar Pulsations), written by Charles Whitney, illustrates stellar pulsation by simulating the thermo-mechanical behavior of a "star" modelled by a self-gravitating gas divided by spherical elastic shells. The elastic shells resemble a set of coupled oscillators. The program solves for the modes of small-amplitude motion, and it uses Fourier synthesis to construct motions for arbitrary starting conditions. The screen displays the thermodynamic structure and surface properties, such as temperature, pressure, and velocity. Animation displays the nature of the pulsation. By showing the motions, temperatures, and energy flux, the program demonstrates the heat engine acting inside the pulsating star. The motions of the shells and the spatial Fourier decomposition into eigenmodes are displayed simultaneously, and this will help you visualize the meaning of the Fourier components.

CLASSICAL MECHANICS PROGRAMS

GENMOT (The Motion Generator), written by Randall Jones, allows you to solve numerically any differential equation of motion for a system with up to three degrees of freedom and display the time evolution of the system in a wide variety of formats. Any of the dynamical variables or any function of those variables may be displayed graphically and/or numerically and a wide range of animations may be constructed. Since the Motion Generator can be used to solve any second-order differential equation, it can also be used to study systems analyzed by Lagrangian methods. Real world coordinates may be constructed as functions of generalized coordinates so that simulations of the actual system can be constructed.

ROTATE (Rotation of 3-D Objects), written by Randall Jones, is designed to aid in the visualization of the dynamical variables of rotational motion. It will allow you to observe the 3-D motion of rotating objects in a controlled fashion, running the simulation faster, slower, or in reverse while displaying the corresponding evolution of the angular velocity, the angular momentum and the torque. It will display the motion from the fixed frame and from the body frame to help in understanding the translation between these two descriptions of the motion. By using the stereographic feature of the program you can create a genuine 3-D representation of the motion of the quantities.

COUPOSC (Coupled Oscillators), written by Randall Jones, is designed to investigate a wide range of harmonic systems. Given a set of objects and springs connected in one or two dimensions, the simulation can solve the problem by generating the normal mode frequencies and their corresponding motions. It can take any set of initial conditions and resolve them into their component normal mode motions or take any set of initial mode occupations and display the corresponding motions of the objects. It can also determine the motion of the system when it is acted on by external forces. In this case the total forces are no longer harmonic, so the solution is generated numerically. The harmonic analysis, however, still provides an important tool for investigating and understanding the subsequent motion.

ANHARM (Anharmonic Oscillators), written by Bruce Hawkins, simulates oscillations of various types: pendulum, simple harmonic oscillator, asymmetric, cubic, Vanderpol, and a mass in the center of a spring with fixed ends. Nonlinear behavior is emphasized. The user may choose to view one to four graphs of the motion simultaneously, along with the potential diagram and a picture of the moving object. Graphs that may be viewed are x vs. t, v vs. t, v vs. x, the Poincaré diagram, and the return map. Tools are provided to explore parameter space for regions of interest. Fourier analysis is available, resonance diagrams can be plotted, and the period can be plotted as a function of energy. Includes a tutorial demonstrating the usefulness of phase plots and Poincaré plots.

ORBITER (Gravitational Orbits), written by Bruce Hawkins, simulates the motion of up to five objects with mutually gravitational attraction, and any reasonable number of additional objects moving in the gravitation field of the first five. The motion may be viewed in up to six windows simultaneously: windows centered on a particular body, on the center of mass, stationary in the universe frame, or rotating with the line joining the two most massive bodies. A menu of available system includes the solar system, the sun/earth/moon system; the sun, Jupiter, and its moons; the sun, earth, and Saturn, demonstrating retrograde motion; the sun, Jupiter, and a comet; and a pair of binary stars with a comet. Bodies my be added to any system, or a new system created using either numerical coordinates or the mouse. Bodies may be replicated to demonstrate the sensitivity of orbits to initial conditions.

COLISION (Collisions), written by Bruce Hawkins, simulates two-body collisions under any of a number of force laws: Coulomb with and without shielding and truncation, hard sphere, soft sphere (harmonic), Yukawa, and Woods-Saxon. Collision may be viewed in the laboratory and center of mass systems simultaneously, with or without momentum diagrams. Includes a tutorial on the usefulness of the center of mass system, one on the kinematics of relativistic collisions, and one on cross section. Plots cross section against scattering parameter, and compares collisions at different parameters.

ELECTRICITY AND MAGNETISM PROGRAMS

FIELDS (Analysis of Vector and Scalar Fields), written by Jarek Tuszynski, displays scalar and vector fields for any algebraic or trigonometric expression entered by the user. It also computes numerically the divergence, curl, and Laplacian for the vector fields, and the gradient and Laplacian for the scalar fields. Simultaneous displays of selected quantities are provided in user-selected planes, using vector, contour, or 3-D plots. The program also allows the user to define paths along which line integrals are computed.

GAUSS (Gauss' Law), written by Jarek Tuszynski, treats continuous charge distributions having spherical or cylindrical symmetry, and those that vary as a function of the x-coordinate only. The program allows the user to enter an arbitrary function to define either the electric field magnitude, the potential, or the charge density. It then computes the other two functions by numerical differentiation or integration, and displays all three functions. Finally, the program allows the user to enter a "comparison function," which is plotted on the same graph, so as to check whether his analytic solutions are correct.

POISSON (Poisson's Equation Solved on a Grid), written by Jarek Tuszynski, solves Poisson's equation iteratively on a 2-D grid using the method of simultaneous over-relaxation. The user can draw arbitrary systems consisting of line charges, and charged conducting cylinders, plates, and wires, all infinite in extent perpendicular to the grid. After iteratively solving Poisson's equation, the program displays the results for the potential, electric field, or the charge density (found from the Laplacian of the potential), in the form of contour, vector, or 3-D plots. In addition, many other program features are available, including the ability to specify surfaces, along which the potential varies according to some algebraic function specified by the user.

IMAG&MUL (Image Charges and Multipole Expansion), written by Lyle Roelofs and Nathaniel Johnson, allows users to explore two approaches to the solution of Laplace's equation—the image charge method and expansion in multipole moments. In the image charge mode (IC) the user is presented with a variety of configurations involving conducting planes and point charges and is asked to "solve" each by placing image charges in the appropriate locations. The program displays the electric field due to all point charges, real and image, and a solution can be regarded as successful with the field due to all charges is everywhere orthogonal to all conducting surfaces. Solutions can then be examined with a variety of included software "tools." The multipole expansion (ME) mode of the program also permits a "hands-on" exploration of standard electrostatic problems, in this case the "exterior" problem, i.e., the determination of the field outside a specified equipotential surface. The program presents the user with a variety of azimuthally symmetric equipotential surfaces. The user "solves" for the full potential by adding chosen amounts of the (first six) multipole moments. The screen shows the contours of the summed potential and the problem is "solved" when the innermost contour matches the given equipotential surface as closely as possible.

ATOMPOL (Atomic Polarization), written by Lyle Roelofs and Nathaniel Johnson, is an exploration of the phenomenon of atomic polarization. Up to 36 atoms of controllable polarizibility are immersed in an external electric field. The program solves for and displays the field throughout the region in which the atoms are located. A closeup window shows the polarization of selected atoms and software "tools" allow for further analysis of the resulting electric fields. Use of this program improves the student's understanding of polarization, the interaction of polarized entities and the atomic origin of macroscopic polarization, the latter via study of closely spaced clusters of polarizable atoms.

DIELECT (Dielectric Materials), written by Lyle Roelofs and Nathaniel Johnson, is a simulation of the behavior of linear dielectric materials using a cell-based approach. The user controls either the polarization or the susceptibility of each cell in a (25 × 25) grid (with assume uniformity in the third direction). Full self-consistent solutions are obtained via an iterative relaxation method and the fields P, E, or D are displayed. The student can investigate the self-interactions of polarized materials and many geometrical effects. Use of this program aids the student in developing understanding of the subtle relations among and meaning of P, E, and D.

ACCELQ (Fields From an Accelerated Charge), written by Ronald Stoner, simulates the electromagnetic fields in the plane of motion generated by a point charge that is moving and accelerating in two dimensions. The user chooses from among seven predefined trajectories, and sets the values of maximum speed and viewing time. The electric field pattern is recomputed after each change of trajectory or parameter; thereafter, the user can investigate the electric field, magnetic field, retarded potentials, and Poynting-vector field by using the mouse as a field probe, by using gridded overlays, or by generating plots of the various fields along cuts through the viewing plane.

QANIMATE (Fields From an Accelerated Charge—Animated Version), written by Ronald Stoner, is an interactive animation of the changing electric field pattern generated by a point electric charge moving in two dimensions. Charge motion can be manipulated by the user from the keyboard. The display can include electric field lines, radiation wave fronts, and their points of intersection. The motion of the charge is controlled by the using **arrow** keys to accelerate and steer much like the accelerator and steering wheel of a car, except that acceleration must be changed in increments, and the **Space** bar can used to engage or disengage the steering. With steering engaged, the charge will move in a circle. Unless the acceleration is made zero, the speed will increase (or decrease) to the maximum (minimum) possible value. At constant speed and turning rate, the charge can be controlled by the **Space** bar alone.

EMWAVE (Electromagnetic Waves), written by Ronald Stoner, uses animation to illustrate the behavior of electric and magnetic fields in a polarized plane electromagnetic wave. The user can choose to observe the wave in free space, or to see the effect on the wave of incidence on a material interface, or to see the effects of optical elements that change its polarization. The user can change the polarization state of the incident wave by specifying its Stokes parameters. Standing electromagnetic waves can be simulated by combining the incident travelling wave with a reflected wave of the same amplitude. The user can do that by choosing appropriate values of the physical properties of the medium on which the incident wave impinges in one of the animations.

MAGSTAT (Magnetostatics), written by Ronald Stoner, computes and displays magnetic fields in and near magnetized materials. The materials are uniform and have 3-D shapes that are solids of revolution about a vertical axis. The shape of the material can be modified or chosen from a data input screen. The user has the option of generating the fields produced by a permanently and uniformly magnetized object, or of generating the fields of a magnetizable object placed in an otherwise uniform external field. Besides choosing the shape and aspect ratio of the object, the user can vary the magnetic permeability of the magnetizable material, and choose among three fields to display: magnetic induction (B), magnetic field strength (H), and magnetization (M). Each of these fields can be displayed or explored in several different ways. The algorithm for computing the fields uses a superposition of Chebyschev polynomial approximants to the H field due to "rings" of "magnetic charge."

MODERN PHYSICS PROGRAMS

NUCLEAR (Nuclear Energetics and Nuclear Counting), written by Michael Moloney, deals with basic nuclear properties related to mass, charge, and energy, for approximately 1900 nuclides. Graphs are available involving binding energy, mass, and Q values of a variety of nuclear reactions, including alpha and beta decays. Part 2 deals with simulating the statistics of counting with a Geiger-Muller tube. This part also simulates neutron activation, and the counting behavior as neutron flux is turned on and off. Finally, a decay chain from A to B to C is simulated, where half-lives may be changed, and populations are graphed as a function of time.

GERMER (Davisson-Germer and G. P. Thomson Experiments), written by Michael Moloney, simulates both the Davisson-Germer and G. P. Thomson experiments with electrons scattering from crystalline materials. Stress is laid on the behavior of electrons as waves; similarities are noted with scattering of x-rays. The exercises encourage students to understand why peaks and valleys in scattered electrons occur where they do.

QUANTUM (one-dimensional Quantum Mechanics), written by Douglas Brandt, is a program that has four sections. The first section allows users to investigate the uncertainty principle for specified wavefunctions in position or momentum space. The second section allows users to investigate the time evolution of wavepackets under various dispersion relations. The third section allows users to investigate solutions to Schrödinger's equation for asymptotically free solutions. The user can input a barrier and the program calculates reflection and transmission coefficients for a range of energies and show wavepacket time evolution for the barrier potential. The fourth section is similar to the third, except that it allows the user to investigate bound solutions to Schrödinger's equation. The program calculates the bound state Hamiltonian eigenvalues and spatial eigenfunctions.

RUTHERFD (Rutherford Scattering), written by Douglas Brandt, is a program for investigating classical scattering of particles. A scattering potential can be chosen from a list of predefined potentials or an arbitrary potential can be input by the user. The computer generates scattering events by randomly picking impact parameters from a distribution defined by beam parameters specified by the user. It displays the results of the scattering on a polar histogram and on a detailed histogram to help users gain insight into differential scattering cross section. A scintillation mode can be chosen for users that want more appreciation of the actual experiments of Geiger and Marsden. A "guess the scatterer" mode is available for trying to gain appreciation of how scattering experiments are used to infer properties of the scatterers.

SPECREL (Special Relativity), written by Douglas Brandt, is a program to investigate special relativity. The first section is to investigate change of coordinate systems through Minkowski diagrams. The user can define coordinates of objects in one reference frame and the computer calculates the coordinates in a user-selectable coordinate system and displays the objects in both reference frames. The second section allows users to view clocks that are in relative motion. A clock can be given an arbitrary trajectory through space-time and the readings of various clocks can be viewed as the clock follows that trajectory. A third section allows users to observe collisions in different reference frames that are related by Lorentz transformations or by Gallilean transformations.

LASER (Lasers), written by Michael Moloney, simulates a three-level laser, with the user in control of energy level parameters, temperature, pump power, and end mirror transmission. Atomic populations may be graphically tracked from thermal equilibrium through the lasing threshold. A mirror cavity simulation is available which uses ray tracing. This permits study of cavity stability as a function of mirror shape and position, as well as beam shape characteristics within the cavity.

HATOM (Hydrogenic Atoms), written by John Hiller, computes eigenfunctions and eigenenergies for hydrogen, hydrogenic atoms, and single-electron diatomic ions. Hydrogenic atoms may be exposed to uniform electric and magnetic fields. Spin interactions are not included. The magnetic interaction used is the quadratic Zeeman term; in the absence of spin-orbit coupling, the linear term adds only a trivial energy shift. The unperturbed hydrogenic eigenfunctions are computed directly from the known solutions. When external fields are included, approximate results are obtained from basis-function expansions or from Lanczos diagonalization. In the diatomic case, an effective nuclear potential is recorded for use in calculation of the nuclear binding energy.

NUCLEAR AND PARTICLE PHYSICS PROGRAMS

NUCLEAR (Nuclear Energetics and Counting), written by Michael Moloney, is included here, but is described under the Modern Physics Heating.

SHELLMOD (Nuclear Models), written by Roberta Bigelow, calculates energy levels for spherical and deformed nuclei using the single particle shell model. You can explore how the nuclear potential shape, the spin-orbit interaction, and deformation affect both the order and spacing of nuclear energy levels. In addition, you will learn how to predict spin and parity for single particle states.

NUCRAD (Interaction of Radiation With Matter), written by Roberta Bigelow, is a simulation of alpha particles, muons, electrons, or photons interacting with matter. You will develop an understanding of how ranges, energy losses, and random particle paths depend on materials, radiation, and incident energy. As a specific application, you can explore photon and electron interactions in a sodium iodide crystal which determines the energy response of a radiation detector.

ELSCATT (Electron-Nucleus Scattering), by John Philpott, is an interactive software tool that demonstrates various aspects of electron scattering from nuclei. Specific features include the relativistic kinematics of electron scattering, densities and form factors for elastic and inelastic scattering, and the nuclear Coulomb response. The simulation illustrates how detailed nuclear structure information can be obtained from electron scattering measurements.

TWOBODY (Two-Nucleon Interactions), by John Philpott, is an interactive software tool that illuminates many features of the two-nucleon problem. Bound state wavefunctions and properties can be calculated for a variety of interactions that may include non-central parts. Phase shifts and cross sections for pp, pn, and nn scattering can be calculated and compared with those obtained experimentally. Spin-polarization features of the cross sections can be extensively investigated. The simulation demonstrate the richness of the two-nucleon data and its relation to the underlying nucleon-nucleon interaction.

RELKIN (Relativistic Kinematics), by Joseph Rothberg, is an interactive program to permit you to explore the relativistic kinematics of scattering reactions and two-body particle decays. You may choose from among a large number of initial and final states. The initial momentum of the beam particle and the center of mass angle of a secondary can also be specified. The program displays the final state vector momenta in both the lab system and center of the mass system along with numerical values of the most important kinematic quantities. The program may be run in a Monte Carlo mode, displaying a scatter plot and histogram of selected variables. The particle data base may be modified by the user and additional reactions and decay modes may be added.

DETSIM (Particle Detector Simulation), by Joseph Rothberg, is an interactive tool to allow you to explore methods of determining parameters of a decaying particle or scattering reaction. The program simulates the response of high-energy particle detectors to the final-state particles from scattering or decays. The detector size and location may be specified by the user as well as its energy and spatial resolution. If the program is run in a Monte Carlo mode, detector hit information for each event is written to a file. This file can be read by a small reconstruction and plotting program. You can easily modify one of the example reconstruction programs that are provided to determine the mass, momentum, and other properties of the initial particle or state.

QUANTUM MECHANICS PROGRAMS

BOUND1D (Bound States in One Dimension), written by Ian Johnston, is a tool which allows you to explore energy eigenfunctions for an electron in various potential wells, which can be square, parabolic, ramped, asymmetric, double or Coulombic. The first part of the program deals with finding the eigenvalues and eigenfunctions of different wells. You may find them yourself, using a "hunt and shoot" method, or else the program will compute the eigenvalues automatically, by counting the number of nodes to determine where the eigenvalues occur. The second part of the program looks at properties of eigenfunctions normalization, orthogonality and the evaluation of many kinds of overlap integrals. The third part examines time development of general states made up of a superposition of bound state eigenfunctions. Facility is provided for you to incorporate your own procedures to specify different potential wells or different overlap integrals.

SCATTR1D (Scattering in One Dimension), written by John Hiller, solves the time-independent Schrödinger equation for stationary scattering states in one-dimensional potentials. The wavefunction is displayed in a variety of ways, and the transmission and reflection probabilities are computed. The probabilities may be displayed as functions of energy. The computations are done by numerically integrating the Schrödinger equation from the region of the transmitted wave, where the wavefunction is known up to some overall normalization and phase, to the region of the incident wave. There the reflected and incident waves are separated. The potential is assumed to be zero in the incident region and constant in the transmitted region.

QMTIME (Quantum Mechanical Time Development), written by Daniel Styer, simulates quantal time development in one dimension. A variety of initial wave packets (Gaussian, Lorentzian, etc.) can evolve in time under the influence of a variety of potential energy functions (step, ramp, square well, harmonic oscillator, etc.) with or without an external driving force. A novel visualization technique simultaneously displays the magnitude and phase of complex-valued wave functions. Either position-space or momentum-space wave functions, or both, can be shown. The program is particularly effective in demonstrating the classical limit of quantum mechanics.

LATCE1D (Wavefunctions on a one-dimensional Lattice), written by Ian Johnston, is a tool which allows you to explore energy eigenfunctions for an electron in a lattice made up of a number of simple potential wells (up to twelve), which can be square, parabolic, or Coulombic. You may find the eigenvalues yourself, using a "hunt and shoot" method, or allow the program to compute them automatically. You can firstly explore regular lattices, where all wells are the same and spaced at regular intervals. These will demonstrate many of the properties of regular crystals, particularly the existence of energy bands. Secondly you can change the width, depth or spacing of any of the wells, which will mimic the effect of impurities or other irregularities in a crystal. Lastly you can apply an external electric across the lattice. Facility is provided for you to incorporate your own procedures to calculate wells, lattice arrangements or external fields of their own choosing.

BOUND3D (Bound States in Three Dimensions), written by Ian Johnston, is a tool which allows you to explore energy eigenfunctions for an particle in a spherically symmetric potential well, which can be square, parabolic, Coulombic, or several other shapes of importance in molecular or nuclear applications. The first part of the program deals with finding the eigenvalues and eigenfunctions of different wells, assuming that the angular part of the wavefunctions are spherical harmonics. You may find them yourself for a given angular momentum quantum number using a "hunt and shoot" method, or else the program will compute the eigenvalues automatically, by counting the number of nodes to determine where the eigenvalues occur. The second part of the program looks at properties of eigenfunctions normalization, orthogonality and the evaluation of many kinds of overlap integrals. Facility is provided for you to incorporate your own procedures to specify different potential wells or different overlap integrals.

IDENT (Identical Particles in Quantum Mechanics), written by Daniel Styer, shows the probability density associated with the symmetrized, antisymmetrized, or nonsymmetrized wave functions of two noninteracting particles moving in a one-dimensional infinite square well. It is particularly valuable for demonstrating the effective interaction of noninteracting identical particles due to interchange symmetry requirements.

SCATTR3D (Scattering in Three Dimensions), written by John Hiller, performs a partial-wave analysis of scattering from a spherically symmetric potential. Radial and 3-D wave functions are displayed, as are phase shifts, and differential and total cross sections. The analysis employs an expansion in the natural angular momentum basis for the scattering wavefunction. The radial wavefunctions are computed numerically; outside the region where the potential is important they reduce to a linear combination of Bessel functions which asymptotically differs from the free radial wavefunction by only a phase. Knowledge of these phase shifts for the dominant values of angular momentum is used to approximate the cross sections.

CYLSYM (Cyllindrically Symmetric Potentials), written by John Hiller, solves the time-independent Schrödinger equation Hu=Eu in the case of a cylindrically symmetric potential for the lowest state of a chosen parity and magnetic quantum number. The method of solution is based on evolution in imaginary time, which converges to the state of the lowest energy that has the symmetry of the initial guess. The Alternating Direction Implicit method is used to solve a diffusion equation given by $HU = -\hbar \partial U / \partial t$, where H is the Hamiltonian that appears in the Schrödinger equation. At large times, U is nearly proportional to the lowest eigenfunction of H, and the expectation value $\langle H \rangle = \langle U | H | U \rangle / \langle U | U \rangle$ is an estimate for the associated eigenenergy.

SOLID STATE PHYSICS

LATCE1D (Wavefunctions for a one-dimensional Lattice), written by Ian Johnston, and included here, is described under the Quantum Mechanics heading.

SOLIDLAB (Build Your Own Solid State Devices), written by Steven Spicklemire, is a simulation of a semiconductor device. The device can be "drawn" by the user, and the characteristics of the device adjusted by the user during the simulation. The user can see how charge density, current density, and electric potential vary throughout the device during its operation.

LCAOWORK (Wavefunctions in the LCAO Approximation), written by Steven Spicklemire, is a simulation of the interaction of 2-D atoms within small atomic clusters. The atoms can be adjusted and moved around while their quantum mechanical wavefunctions are calculated in real time. The student can investigatge the dependence of various properties of these atomic clusters on the properties of individual atoms, and the geometric arrangement of the atoms within the cluster.

PHONON (Phonons and Density of States), written by Graham Keeler, calculates and displays phonon dispersion curves and the density of states for a number of different 3-D cubic crystal structures. The displays of the dispersion curves show realistic curves and allow the user to study the effect of changing the interatomic forces between nearest and further neighbor atoms and, for diatomic crystal structures, changing the ratio of the atomic masses. The density of states calculation shows how the complex shapes of real densities of states are built up from simpler distributions for each mode of polarization, and enables the user to match the features of the distribution to corresponding features on the dispersion curves. In order to help with visualization of the crystal lattices involved, the program also shows 3-D projections of the different crystal structures.

SPHEAT (Calculations of Specific Heat), written by Graham Keeler, calculates and displays the temperature variation of the lattice specific heat for a number of different theoretical models, including the Einstein model and the Debye model. It also makes the calculation for a computer simulation of a realistic density of states, in which the user can vary the important parameters of the crystal, including those affecting the density of state. The program can display the results for a small region near the origin, and as a T-cubed plot to enable the user to investigae the low temperature limit of the specific heat, or in the form of the equivalent Debye temperature to enhance a study of the deviations from the Debye model. The Schottky specific heat anomaly can also be investigated.

BANDS (Energy Bands), written by Roger Rollins, calculates and displays, for easy comparison, the energy dispersion curves and corresponding wavefunctions for an electron in a 1-D symmetric $V(x) = V(-x)$ periodic potential of arbitrary shape and of strength V_0. The method used is based on an exact, non-perturbative approach so that the energy dispersion curves and band gaps can be obtained for large V_0. Wavefunctions can be displayed, and compared with one another, by clicking the mouse on the desired states on the energy dispersion curve. Changes in band strtucture can be followed as changes are made in the shape of the potential. The variation of the band gaps with V_0 is calculated and compared with the two opposite limits of very weak V_0 (perturbation method) and very strong V_0 (isolated atom). Even the experienced condensed matter researcher may be surprised by some of the results! Open-ended class discussions can result from the interesting physics found in these conceptually simple model calculations.

PACKET (Electron Wavepacket in a 1-D Lattice), written by Roger Rollins, shows a live animation, calculated in real time, demonstrating how an electron wavepacket in a metal or semiconducting crystal moves under the influence of external forces. The time-dependent Schrödinger equation is solved in a tight binding approximation, including the external force terms, and the motion of the wavepacket is obtained directly. The main objective of the simulation is to show that an electron wavepacket formed from states with energies near the top of an energy band is accelerated in a direction *opposite* to the direction of the external force; it has a *negative* effective mass! The simulation deals with motion in a 1-D lattice but the concepts are applicable to the full 3-D motion of an electron in a real crystal. Numerical experiments on the motion of the packet explore interesting physics questions such as: how does constant applied force affect the periodic motion of a packet? when does the usual semiclassical model fail? what happens to the dynamics of the packet when placed in a superlattice with lattice constant twice that of the original lattice?

THERMAL AND STATISTICAL PHYSICS PROGRAMS

ENGDRV, written by Lynna Spornick, is a driver program for **ENGINE, DIESEL, OTTO, and WANKEL**. These programs provide an introduction to the thermodynamics of engines.

ENGINE (Design Your Own Engine), written by Lynna Spornick, lets the user design an engine by specifying the processes (adiabatic, isobaric, isochoric [constant volume], and isothermic) in the engine's cycle, the engine type (reversible or irreversible), and the gas type (helium, argon, nitrogen, or steam). The thermodynamic properties (heat exchanged, work done, and change in internal energy) for each process and the engine's efficiency are computed.

DIESEL, OTTO, and WANKEL, written by Lynna Spornick, provide animations of each of these types of engine. Plots of the temperature versus entropy and the pressure versus volume for the cycles are show with the engine's current thermodynamic conditions indicated.

PROBDRV, written by Lynna Spornick, is a driver program for **GALTON, POISEXP, TWOD, KAC, and STADIUM**. Subprograms GALTON, POISEXP, and TWOD provide an introduction to probability and subprograms KAC and STADIUM provide an introduction to statistics.

GALTON (A Galton Board), written by Lynna Spornick, models either a traditional Galton Board or a customized Galton Board with traps, reflecting, and/or absorbing walls. GALTON demonstrates the binominal and normal distributions, the laws of probability, and the central limit theorem.

POISEXP (Poisson Probability Distribution in Nuclear Decay), written by Lynna Spornick, uses the decay of radioactive atoms to describe the Poisson and the exponential distributions.

TWOD (2-D Random Walk), written by Lynna Spornick, models a random walk in two dimensions. A "drunk," taking equal-length steps, is required to walk either on a grid or on a plane. TWOD demonstrates the joint probability of two independent processes, the binominal distribution, and the Rayleigh distribution.

KAC (A Kac Ring), written by Lynna Spornick, uses a Kac ring to demonstrate that large mechanical systems, whose equations of motion are solvable and which obey time reversal and have a Poincaré cycle, can also be described by statistical models.

STADIUM (The Stadium Model), written by Lynna Spornick, uses a stadium model to demonstrate that there exists mechanical systems whose equations of motion are solvable but whose motion is not predictable because of the system's chaotic nature.

ISING (Ising Model in One and Two Dimensions), written by Harvey Gould, allows the user to explore the static and dynamic properties of the 1- and 2-D Ising model using four different Monte Carlo algorithms and three different ensembles. The choice of the Metropolis algorithm allows the user to study the Ising model at constant temperature and external magnetic field. The orientation of the spins is shown on the screen as well as the evolution of the total energy or magnetization. The mean energy, magnetization, heat capacity, and susceptibility are monitored as a function of the number of configurations that are sampled. Other computed quantities include the equilibrium-averaged energy and magnetization autocorrelation functions and the energy histogram. Important physical concepts that can be studied with the aid of the program include the Boltzmann probability, the qualitative behavior of systems near critical points, critical exponents, the renormalization group, and critical slowing down. Other algorithms that can be chosen by the user correspond to spin exchange dynamics (constant magnetization), constant energy (the demon algorithm), and single cluster Wolff dynamics. The latter is particularly useful for generating equilibrium configurations at the critical point.

MANYPART (Many Particle Molecular Dynamics), written by Harvey Gould, allows the user to simulate a dense gas, liquid, or solid in two dimensions using either molecular dynamics (constant energy, constant volume) or Monte Carlo (constant temperature, constant volume) methods. Both hard disks and the Lennard-Jones interaction can be chosen. The trajectories of the particles are shown as the system evolves. Physical quantities of interest that are monitored include the pressure, temperature, heat capacity, mean square displacement, distribution of the speeds and velocities, and the pair correlation function. Important physical concepts that can be studied with the aid of the program include the Maxwell-Boltzmann probability distribution, fluctuations, equation of state, correlations, and the importance of chaotic mixing.

FLUIDS (Thermodynamics of Fluids), written by Jan Tobochnik, allows the user to explore the fluid (gas and liquid) phase diagrams for the van der Waals model and water. The user chooses four phase diagrams from among the following choices: *PT, Pv, vT, uT, sT, uv,* and *sv,* where P is the pressure, T is the temperature, v is the specific volume, S is the specific entropy, and u is the specific internal energy. The program reads in the coexistence table for the van der Waals model and water, and uses it along with an empirical formula for the water free energy and the free energy derived from the van der Waals model. Given v and u, any thermodynamic quantity can be calculated. For the van der Waals model thermodynamic quantities also can be calculated from the other thermodynamic state variables. The user can draw a straight line path in one phase diagram and see how this path looks in the other phase diagrams. The user also can extract all important thermodynamic data at any point in a phase diagram.

QMGAS1 (Quantum Mechanical Gas—Part 1), written by Jan Tobochnik, does the numerical calculations necessary to solve for the thermodynamic properties of quantum ideal gases, including photons in blackbody radiation, ideal bosons, phonons in the Debye theory, non-interacting fermions, and the classical limits of these systems. The user chooses the type of statistics (Bose-Einstein, Fermi-Dirac, or Maxwell-Boltzmann), the dimension of space, the form of the dispersion relation (restricted to simple powers), whether or not the particles have a non-zero chemical potential, and whether or not there is a Debye cutoff. The program then allows the user to build up a table of thermodynamic data, including the energy, specific heat, and chemical potential as a function of temperature. This data and various distribution functions and the density of states can be plotted.

QMGAS2 (Quantum Mechanical Gas—Part2), written by Jan Tobochnik, implements a Monte Carlo simulation of a finite number of quantum particles fluctuating between various states in a finite k-space (k is the wavevector). The program orders the possible energy states into an energy level diagram and then allows particles to move from one state to another according to the usual Boltzman probability distribution. Bosons are restricted so that they may not pass through each other on the energy level diagram; fermions are further restricted so that no two fermions may be in the same state; classical particles have no restrictions. In this way indistinguishability is correctly implemented for bosons and fermions. The user chooses the type of particle, the number of particles, the size and dimension of k-space, and the temperature. During the simulation the user sees a representation of the state occupancy and plots of the average energy, the instantaneous energy, and the distribution of energy among the states, also shown are results for the average energy, specific heat, and the occupancy of the ground state.

WAVES AND OPTICS PROGRAMS

DIFFRACT (Interference and Diffraction), by Robin Giles, simulates some of the fundamental wave behaviors in Fresnel and Fraunhofer Diffraction, and other Interference and Coherence effects. In particular you will be able to study diffraction phenomena associated with a point or a set of points and a slit or set of slits using the Huyghens construction. You can also use a method developed by Cornu—the Cornu Spiral—to examine diffraction from one or two slits or one or two obstacles. You can study Fresnel and Fraunhofer diffraction with a single slit or set of slits, a rectangular aperture and a circular aperture. Finally you can study Partial Coherence and fringe visibility in interference and diffraction observations. In the latter example you will be able to study the Michelson Stellar Interferometer and measure the separation distance in a double star and measure the diameter of single stars.

SPECTRUM (Applications of Interference and Diffraction), by Robin Giles, simulates the uses and modes of operation of four important optical instruments—the Diffraction Grating, the Prism Spectrometer, the Michelson Interferometer and the Fabry-Perot Interferometer. You will look at the nature of the spectra, simulated interference patterns, and the question of resolving power.

WAVE (One-Dimensional Waves), by Wolfgang Christian, Andrew Antonelli, and Susan Fischer, uses finite difference methods to study the time evolution of the following partial differential equations: classical wave, Schrödinger, diffusion, Klein-Gordon, sine-Gordon, phi four, and double sine-Gordon. The user may vary the initial function and boundary conditions. Unique features of the program include mouse-driven drawing tools that enable the user to create sources, segments, and detectors anywhere inside the medium. Double-clicking on a segment allows the user to edit properties such as index of refraction or potential in order to model barrier problems such as thin film interference filters or the Ramsauer-Townsend effect in optics and quantum mechanics, respectively. Various types of analysis can be performed, including detector value, space-time, Fourier analysis and energy density.

CHAIN (One-Dimensional Lattice of Coupled Oscillators), by Wolfgang Christian, Andrew Antonelli, and Susan Fischer, allows the user to examine the time evolution of a 1-D lattice of coupled oscillators. These oscillators are represented on screen as a chain of masses undergoing vertical displacement. The program allows the user to examine how the application of Newtonian mechanics to these masses leads to traveling and standing waves. The relationship between the lattice spacing and other properties such as dispersion, band gaps, and cut-off frequency can be examined. Each mass can be assigned linear, quadratic, and cubic nearest neighbor interactions as well as a time-dependent external force function. Global properties such as the total energy in the lattice or the Fourier transform of the lattice can be displayed as well as the time evolution of a single mass's dynamical variables.

FOURIER (Fourier Analysis and Synthesis), written by Brian James, allows investigation of Fourier analysis and 1-D and 2-D Fourier transforms. In Fourier analysis users can choose from several predefined functions or enter their own functions either algebraically, numerically or graphically. The build-up of a periodic function is illustrated as successive terms of the Fourier series are added in, and the effects of dispersion and attenuation on the evolution of the synthesized waveform can then be investigated. One- and two-dimensional discrete Fourier transforms can be produced for a range of standard and user-entered functions. The effects of filters on the inverse transforms are illustrated. The 2-D transforms are shown as surface and contour plots. Image processing can be illustrated by filtering the transforms of gray level images so that when the inverse transforms are displayed it can be seen that the images have been modified.

RAYTRACE (Ray Tracing and Lenses), by Brian James, lets the user explore the applications of ray tracing in geometrical optics. The fundamental principle of Fermat can be illustrated by plotting the path of a ray through two different materials between fixed points. The variation of the path of a ray through a region of changing refractive index can be used to investigate the formation of mirages. The variation of pulse delay in a fiber can be calculated as a function of its parameters and the characteristics of optical communication fibers are considered. The formation of primary and secondary rainbows due to dispersion of refractive index can be displayed. The matrix method of tracing rays through lenses can be used to investigate the images formed and show how aberrations in images arise and may be reduced.

QUICKRAY (Quick Ray Tracing), by John Philpott, can be used to demonstrate ray diagrams for a single thin lens or spherical mirror. The object and image are shown, along with the three principal rays that proceed from the object towards the observer. You can use the mouse to move the object, the position of the lens or mirror or to change the focal length of the lens or mirror. The principal rays are continuously redrawn while any of these adjustments are made. The simulation handles converging and diverging lenses and concave and convex mirrors. Thus students can quickly get an intuitive feel for real and virtual image formation under a variety of circumstances.

Acknowledgments

The CUPS Project was funded by the National Science Foundation (under grant number PHY-9014548), and it has received support from the IBM Corporation, the Apple Corporation, and George Mason University.

2

Historic Experiments: Rutherford Scattering

Douglas E. Brandt

2.1 Introduction

One of the most common experimental methods used to study the structure of matter is scattering. A scattering experiment, in general, starts with a known incoming beam of particles and a target with which to interact. The incoming beam is allowed to interact with the target and any outgoing particles from the interaction are detected. Any information gathered about the outgoing particles is used to infer properties of the incoming beam, the target, and the interaction between the incoming beam and the target.

As an example of scattering at its most commonplace, most of human vision relies on scattering processes. Most objects are viewed by the light they scatter that reaches their surfaces. The incoming beam is the light that reaches objects' surfaces from the light sources present in the surroundings, the target is the object being viewed, and the outgoing particles are the photons scattered off the object. The detector is the human eye. The human brain makes many judgments about the nature of objects it views based on the data received by the eye such as size and shape, texture, and composition.

As an example of scattering that involves the limits of current technology, almost all current experiments in high-energy physics involve measuring the scattering of one particle from another. Particles are accelerated to energies many times their rest mass to collide with other particles. The results of the collision are detected by sophisticated equipment. The data from these experiments is used to determine properties of the structure of matter at the most fundamental levels yet investigated.

Early important investigations of matter through scattering experiments were carried out by Ernest Rutherford and his colleagues. In Rutherford's experiment, alpha particles emitted by daughters of radium nuclei were scattered off of thin materials. Of particular significance was the approximately 8 MeV alpha decay of ^{214}Po. Rutherford and his colleagues' results were used to determine the basic structure of the modern model of the atom with a tiny positively charge massive nucleus and a relatively low density cloud of negative charge surrounding it.

The program SCATTER allows computer investigation of problems similar to the one that Rutherford studied in his laboratory. What purpose does this simulation serve when one could go into a laboratory and attempt to repeat Rutherford's experiment? This program allows one to change the interaction between the incoming beam and the target, which is something that cannot be done in the laboratory. Hopefully this will help the student develop an intuition about scattering results and gain an understanding of the importance of scattering experiments in the development of modern physics. As an example, one of the exercises is to replace the known structure of atoms with the "plum pudding" model of the structure of matter. This was a model of matter contemporary with Rutherford's experiment which was rejected as a result of Rutherford's investigation.

The program SCATTER is limited to classical scattering problems, which means that the wave nature of the particles colliding is not considered in determining the result of the scattering process. The scattering of alpha particles in Rutherford scattering can be understood without introducing the wave nature of matter just as most visual phenomena can be explained by geometric optics. For an investigation of scattering that treats the wave nature of matter, see the *CUPS Quantum Mechanics*[1] simulations SCATTR1D and SCATTR3D.

2.2   Rutherford's Experiment and the Model of the Atom

Ernest Rutherford recognized that energetic alpha particles could be used to probe the internal structure of matter in 1906[2] when he conducted an experiment in which alpha particles passed through a piece of mica. He observed that most of the alpha particles were deflected by up to two degrees from their initial paths. Simple calculations of the electrostatic forces on the alpha particles led Rutherford to realize that very intense electric fields must exist inside matter to cause this deflection.

Geiger and Marsden[3] continued the study of the deflection of alpha particles passing through thin sheets of materials. They found that there is some small probability of a great angle of deflection (greater than 90 degrees) as the alpha particle passes through matter. In fact, it was Marsden (an undergraduate student at the time!) who first observed the backscattering of the alpha particles. The distribution in angle of the scattered alpha particles led Rutherford to reject the explanation that the alpha particles scattering through large angles was the result of many small angle scatterings and to conclude that the large angle scattering was the result of a single scattering event.

In 1881, J. J. Thomson[4] proposed an early model of the atom. His model consisted of a number of electrons and an equal amount of positive charge distributed uniformly throughout the atom. This model is often referred to as the "plum pudding" model of the atom in text books. In 1910, Thomson published a paper using his atomic model to explain the deflection of charged particles passing through matter. The model was tested experimentally by Crowther[5] using beta particles. Crowther's results were in agreement with Thomson's model. However, the results of Geiger and Marsden demonstrating large angle deflection were not in agreement. Rutherford[6] realized that Thomson's model would need to be altered to agree with the large angle scattering observed for a single event. If all of the positive charge was concentrated at one location within the atom rather than uniformly distributed, the predicted deflection of alpha particles would be in agreement with experimental evidence. This was the origin of the modern nuclear model of the atom.

2.3 Differential Cross Section

The connection between theory and experiment in scattering is made through the differential cross section for scattering. Theory, in principle, can predict the differential cross section for the interaction between the scatterers. A scattering experiment attempts to measure the differential cross section defined below. A comparison of the theoretical and experimental results can be made to test a particular model of the interaction between the scatterers.

The differential cross section is defined as

$$d\sigma = \frac{\text{number of particles per unit time into a solid angle } d\Omega}{\text{incident flux of particles}}, \qquad (2.1)$$

where $d\sigma(\theta, \phi)$ is the differential cross section for scattering at angles θ and ϕ; $d\Omega(\theta, \phi)$ is a differential solid angle in the direction given by θ and ϕ; and the incident flux of particles is the number of particles per unit area per unit time in the incident particle beam. In situations that can be treated with classical mechanics, (Fig. 2.1) this can be thought of as the differential cross-sectional area of the incident beam that gets scattered into the range of directions bounded by the polar angles θ and $\theta + d\theta$ and the azimuthal angles ϕ and $\phi + d\phi$.

In an experiment, only the number of particles per unit time into a finite solid angle can be measured, not into an infinitesimal solid angle. All detectors have a finite area of detection, A, which implies they subtend a finite solid angle at finite distances,

$$\Delta\Omega = \frac{A}{4\pi r^2}, \qquad (2.2)$$

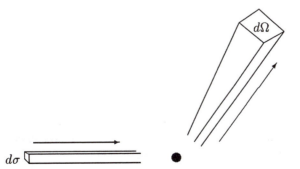

Figure 2.1: In classical scattering, the differential cross section, $d\sigma$, is the area of the incident beam that scatters into the solid angle $d\Omega$.

where $\Delta\Omega$ is the finite solid angle subtended by the detector and r is the distance from the target to the detector. The number of particles per unit time into the detector is related to the differential cross section by

$$\Delta\sigma = \int_{\Delta\Omega} d\sigma = \int_{\Delta\Omega} \frac{d\sigma}{d\Omega} d\Omega. \tag{2.3}$$

If $d\sigma$ is not a rapidly changing function of θ or ϕ over the range of θ and ϕ covered by the detector, then

$$\frac{\Delta\sigma}{\Delta\Omega} \approx \frac{d\sigma}{d\Omega} = \frac{\text{number of particles per unit time into detector}}{\text{incident flux of particles}}. \tag{2.4}$$

In some cases, and most that are illustrated in this program, the scattering potential has azimuthal symmetry. This implies that there will be no dependence of the differential cross section on the azimuthal direction ϕ. Then $d\sigma$ is a function of θ only and the problem is essentially two-dimensional. In an experiment, no new information is found by changing the azimuthal angle of the detector when there is azimuthal or cylindrical symmetry in the scattering potential.

2.4   *Numerical Approach*

The purpose of this section is to introduce the reader to the type of numerical method used in this program to determine the trajectory and hence the scattering angle of the incident particles as they are scattered. It does not contain details about the Runge-Kutta-Fehlberg method that is used by the program, but introduces finite-difference methods of solving initial value problems of ordinary differential equations.

The only numerical method required in this program is a method of integrating the equation of motion of the incident particle beam under the influence of the scattering force. The incident particle starts at an arbitrary distance from the scatterer with an initial velocity and impact parameter. Its trajectory is then determined by Newton's second law,

$$\frac{d^2\vec{r}}{dt^2} = \frac{\vec{F}}{m}. \tag{2.5}$$

This is a second-order ordinary differential equation with position as the dependent variable and time as the independent variable. It is called an initial value problem or single-point boundary value problem because the values of the initial position and velocity are given.

There are several approaches that can be taken at this point. If the force is a central force, the problem simplifies somewhat.[7,8] To allow the most general form of force, it will not be assumed that the force is a central force. The problem will be rewritten in Cartesian coordinates:

$$\frac{d^2x}{dt^2} = \frac{F_x}{m} \tag{2.6}$$

and

$$\frac{d^2y}{dt^2} = \frac{F_y}{m}.$$ (2.7)

Each second-order differential equation can be written in terms of two first-order differential equations:

$$\frac{dv_x}{dt} = \frac{F_x(x, y)}{m}$$ (2.8)

$$\frac{dx}{dt} = v_x$$ (2.9)

$$\frac{dv_y}{dt} = \frac{F_y(x, y)}{m}$$ (2.10)

$$\frac{dy}{dt} = v_y.$$ (2.11)

The set of two coupled second-order equations has been changed into an equivalent set of four first-order coupled equations. The initial values of the dependent variables x, v_x, y, and v_y are given at time $t = 0$.

These equations can be numerically integrated to determine the time evolution of the dependent variables. Euler's method can be used to demonstrate the general idea used to integrate these equations. The discrete equations related to the set of Eq. 2.11 are

$$\Delta v_x = \frac{F_x(x, y)}{m}\Delta t$$ (2.12)

$$\Delta x = v_x \Delta t$$ (2.13)

$$\Delta v_y = \frac{F_y(x, y)}{m}\Delta t$$ (2.14)

$$\Delta y = v_y \Delta t,$$ (2.15)

where Δt is the discrete time step and Δx, Δy, Δv_x, and Δv_y are the discrete changes in the components of position and velocity, respectively, during the time interval Δt. The most naive method to evaluate the changes in the dependent variables during a time step is to use the value of the variables at the beginning of the time interval to evaluate the coefficients of the Δt terms on the right-hand sides of Eq. 2.15 to generate values of the changes in the dependent variables. The dependent variables are then changed to the new values by adding those generated changes, and the independent variable t is changed to $t + \Delta t$, as illustrated in Figure 2.2. The new values of the dependent variables are then used to calculate the changes for the next time step, and the process is repeated until the desired total change in the independent variable has been made.

Euler's method is not very efficient and there are stability problems in some physically reasonable cases.[9] Euler's method is approximating the dependent variables as linear functions of the independent variable during the time step Δt. Consider the relatively simple analytic problem of motion under a constant force from rest. The velocities are linear functions of the independent variable t, but the method

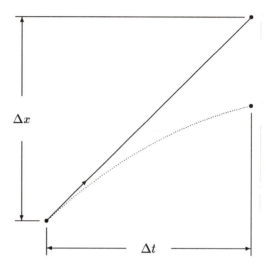

Figure 2.2: Illustration of the Euler method. The velocity at the initial position is used to predict the final position a time Δt later.

of calculating the changes in position will always underestimate the change in position. From introductory physics, the analytic solutions to constant force motion are given by

$$\Delta v_x = v_x - v_0 = a_x \Delta t \qquad (2.16)$$

$$\Delta x = x - x_0 = \frac{1}{2} a_x (\Delta t)^2 + v_0 \Delta t, \qquad (2.17)$$

where x_0 is the position at the beginning of the time step and v_0 is the velocity at the beginning of the time step. Comparing the analytic solutions in Eq. 2.17 to those in Eq. 2.15, it is seen the velocity approximation is exact but that the error in position, ϵx, at the end of the time interval is

$$\epsilon x = \frac{1}{2} a_x (\Delta t)^2. \qquad (2.18)$$

This implies in general that the error that is made in a single step will be of the order of the square of the size of the time step. This means that each time the size of Δt is cut in half, the error in Δx will be approximately four times smaller. That sounds wonderful, but what does it mean about the size of Δt used in a calculation? Should Δt just be made arbitrarily small? Two different problems arise from this approach. The most obvious one is that each time the time step size is cut in half, there will be twice as many calculations to perform to calculate the complete trajectory. Therefore calculations will take twice as long to perform. If time to perform the calculations is not a concern, the finite precision of numbers used by a computer will be. The computer rounds off the result of computations to the nearest number in the precision in which it stores numbers. If the step sizes become too small, the round-off error will contribute a significant additional error to the error generated by the numerical approximation.

Perhaps there is a better way to generate the steps. Can the reader see a better strategy to approaching the integration of the differential equation? With some thought and also knowing that

$$\Delta x = v_{ave}\Delta t \qquad\qquad (2.19)$$

is the definition of average velocity, one may be led to the midpoint method of performing one time step integration of the equations. The average velocity over the interval is unknown, but in many cases the velocity at half a time step may be a reasonable approximation to the average velocity over the interval. It would be exact in the constant force motion illustrated above. The midpoint method (Fig. 2.3) uses a half time step to approximate the average velocity with the predicted velocity at the middle of the time interval and uses this midpoint velocity to predict the change in position for the entire time interval

$$v_{ave} \approx v_{half-step} = v_0 + \frac{F(x_0)}{m}\frac{\Delta t}{2} \qquad\qquad (2.20)$$

$$\Delta x = v_{half-step}\Delta t. \qquad\qquad (2.21)$$

This is illustrated in Figure 2.3.

Where will any error enter into this method? Errors arise from the midpoint method because in a general problem the acceleration changes with position. The prediction of the velocity will be incorrect and the average velocity will not necessarily be the midpoint velocity. However, these errors will be of order of $(\Delta t)^3$. This means that each time Δt is cut in half, there will be approximately a factor eight decrease in the error in a single step. The cost for the increased accuracy is an extra evaluation of the velocity for each time step. It should be easy to see that, because the error size depends on the inverse cube of the step size, this method will have greater accuracy for a given number of computations than will Euler's method for some step size or smaller.

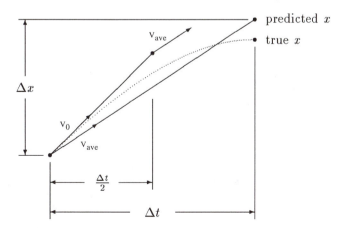

Figure 2.3: Illustration of the midpoint method. The approximate average velocity is calculated by assuming the force at the initial position acts for one-half the time interval. This average velocity is used to calculate the change in position.

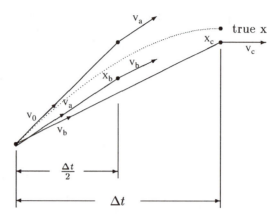

Figure 2.4: Illustration of the fourth-order Runge-Kutta method. First, intermediate velocity approximations are made.

The midpoint method is specific example of a general class of methods of making one time step called Runge-Kutta methods. The midpoint method is called the second-order Runge-Kutta method. It makes the prediction of x at the end of the interval correct to order $(\Delta t)^2$. Higher-order methods can correct to a higher order in Δt, but at a cost of needing to evaluate more functions. The most commonly used Runge-Kutta method is the fourth-order method. The time step is made with the following set of equations:

$$x_a = x_0 + v_0 \frac{\Delta t}{2} \tag{2.22}$$

$$v_a = v_0 + \frac{F(x_0)}{m} \frac{\Delta t}{2} \tag{2.23}$$

$$x_b = x_0 + v_a \frac{\Delta t}{2} \tag{2.24}$$

$$v_b = v_0 + \frac{F(x_a)}{m} \frac{\Delta t}{2} \tag{2.25}$$

$$x_c = x_0 + v_b \Delta t \tag{2.26}$$

$$v_c = v_0 + \frac{F(x_b)}{m} \Delta t \tag{2.27}$$

$$\Delta x = \left(\frac{v_0}{6} + \frac{v_a}{3} + \frac{v_b}{3} + \frac{v_c}{6} \right) \Delta t \tag{2.28}$$

$$\Delta v = \left(\frac{F(x_0)}{6m} + \frac{F(x_a)}{3m} + \frac{F(x_b)}{3m} + \frac{F(x_c)}{6m} \right) \Delta t, \tag{2.29}$$

where x_a is the position prediction for a half-step using the initial velocity, v_a is the velocity prediction using the force at the initial position, x_b is the position prediction for a half-step using v_a for the velocity, v_b is the velocity prediction using the force at x_a, x_c is the position prediction for a whole time step using v_b as the velocity, and v_c is the predicted velocity using the force at x_b. The final change in position, Δx, is determined using a weighted average of the four velocities above. This is illustrated in Figures 2.4 and 2.5. This method makes predictions correct to an order of $(\Delta t)^4$ and so the error can be expected to be of order $(\Delta t)^5$.

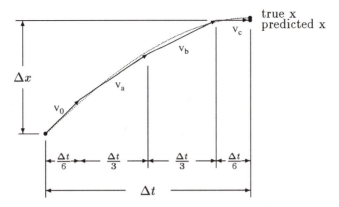

Figure 2.5: Illustration of the fourth-order Runge-Kutta method. A weighted average of the intermediate velocity approximations is used to approximate the final position.

In a region where the acceleration is zero or relatively constant, a great amount of wasted computation will be performed by using small steps with a high-order method. If there is a way to estimate the accuracy of a single step from the data, step size could be adjusted so that big time steps could be used when the acceleration isn't changing rapidly, and the steps could be made smaller when the acceleration is changing rapidly when necessary to retain the desired accuracy in the solution. Such a method is used in this program. The method is called the Runge-Kutta-Fehlberg method.[10] This method compares a fourth-order and a fifth-order Runge-Kutta approximation to estimate the error in a single step and to adjust the step size to keep the error in a single step within the desired limit. Refer to Butcher et al[11] for a detailed discussion of this method.

This discussion would be incomplete without a warning. The Runge-Kutta method can work very well, but there are cases for which it can be very inaccurate (see Exercise 2.14). The reason that it has been used quite often is that it is accurate for the types of problems of interest to many investigators. This program applies the method to forces that are well behaved. However, there is a section of the program that allows users to input their own force. Beware that there are possible problems if the force that is used is not "smooth" everywhere on the scale of the step size used. Because an adaptive step size algorithm is used, the program may be able to handle some functions that are not smooth on the initial step size scale, but the user should be alert to possible problems when using forces of that nature.

2.5   Details of the SCATTER Program

The program SCATTER is contained in the executable file SCATTER.EXE included on the *CUPS Modern Physics* diskette. Assuming the software has been installed in the current default directory as discussed in Chapter 1, start the program from the DOS command line by entering "scatter."

2.5.1   The Display

There are three different basic displays shown on the screen throughout this program. The first screen, shown in Figure 2.6, displays a polar histogram in the lower left portion of the screen, some information about the incident particle beam and the scatterer at the top left of the screen, some sliders at the top right side of the screen for controlling parameters of the incident beam, and a detail histogram with sliders for controlling its width and center value at the bottom right of the screen. A second screen, shown in Figure 2.7, is a display of a diagram from Geiger and Marsden's paper in 1909 describing the apparatus that was used to first observe large angle scattering of alpha particles. The third basic screen used in this program is shown in Figure 2.8. It displays a scintillation screen and three sliders to control parameters associated with the scintillation simulation. The basic intent of this mode of the program is not for students to sit in front of a screen counting scintillations to collect data, but for students to gain some appreciation of the effort involved in performing this work.

2.5.2   Program Control

The program is controlled by making selections from the menus and function keys and by entering appropriate information requested by the program. There are five main menus, each with several choices:

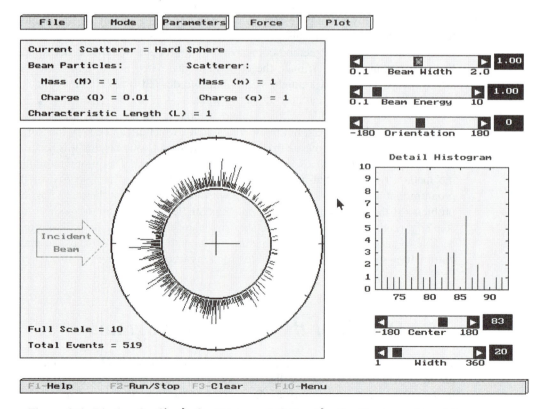

Figure 2.6: Display for **Single Scatterer** and **Guess the Scatterer** modes of the program.

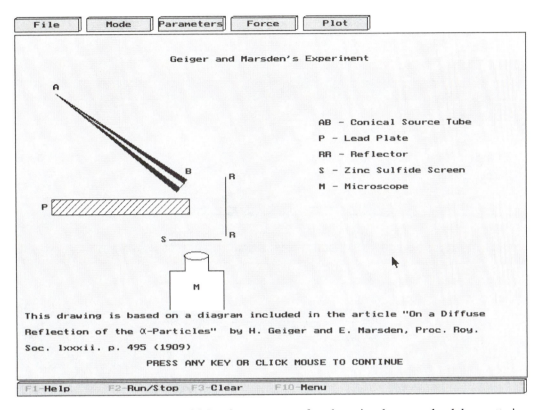

Figure 2.7: Display of Geiger and Marsden apparatus for observing large angle alpha scattering.

- **File:** Program information, configuration, exit program.
  - **About CUPS:** Show description of the CUPS project.
  - **About Program:** Show short description of this program.
  - **About Section:** Show more detailed information about the current section of the program being used.
  - **Configuration:** View and change the configuration settings that can be controlled by the user.
  - **Quit:** Exit the program and return to the operating system.

- **Mode:** Select the mode the program is operating in.
  - **Geiger-Marsden Experiment:** This selection displays a diagram from Geiger and Marsden's 1909 paper describing their large angle alpha scattering experiment. It then allows the user to do a scintillation simulation.
  - **Single Scatterer:** This selection allows the user to generate data for a scattering simulation for various selectable scattering forces and parameters.
  - **Guess the Scatterer:** This selection initiates in a mode in which the program randomly selects a scattering force and the user can perform various scattering experiments to try to determine the nature of the unknown scattering force.

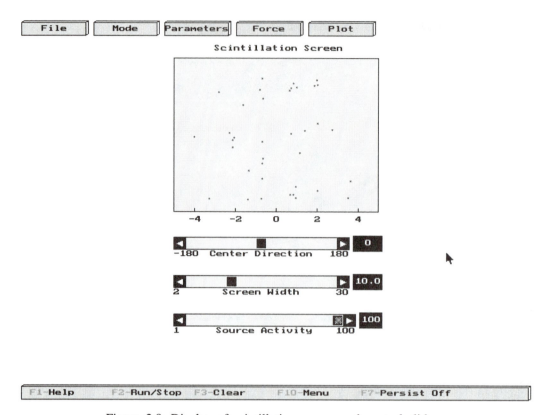

Figure 2.8: Display of scintillation screen and control sliders.

- **Parameters:** Change the parameters of the scattering simulation.

    - **Beam Parameters:** Change the width, divergence, and energy of the incident beam particles.

- **Force:** Select the functional form of the force between the target and beam particles.

    - **Hard sphere:** Selects scattering force to be a hard sphere interaction.
    - **Regular N-Gon:** Selects scattering force to be the interaction between a point particle with a hard, regular n-sided polygon surface.
    - **1/r:** Selects scattering force to be inversely proportional to the distance from the scatterer.
    - $1/r^2$ **Exact:** Selects the scattering force to be inversely proportional to the square of the distance from the scatterer. The scattering is calculated from the exact differential cross section from inverse square scattering.
    - $1/r^2$ **Integrated:** Selects the scattering force to be inversely proportional to the square of the distance from the scatterer. The scattering is calculated from integrating the equation of motion of the beam particles.
    - $1/r^3$**:** Selects the scattering force to be inversely proportional to the cube of the distance between the scatterers.
    - **Yukawa:** Selects the scattering force to be that which arises from the Yukawa potential.

- **Uniform Scatter:** Selects the scattering force to be that which provides uniform scattering into all angles.
  - **User Defined 1: F(r):** Selects the scattering force to be one which the user defines. This force is only a function of the distance from the scatterer.
  - **User Defined 2: F(x, y, vx, vy):** Selects the scattering force to be one which the user defines. This force can be a function of both the coordinates and the velocity of the scatterers.

- **Plot:** This selects options for plotting functions over the detail histogram plot.

  - **Plot Function:** Allows a function to be plotted on the detail histogram for purposes of fitting a function to the data.

- **Hot Keys:** There are a number of hot keys available for controlling the program:

- **F1-Help:** This hot key brings up a context-sensitive help screen.

- **F2-Run/Stop:** This hot key starts and stops the generation and display of the results of scattering events.

- **F3-Clear:** This hot key clears the collected data from the histogram or scintillator data structure and the display.

- **F7-Persist On/Off:** This hot key only functions during the display of the scintillation screen during the Geiger-Marsden experiment mode of the program. It causes the scintillation screen to retain the display of events on the screen as photographic film behaves in the persist-on mode. In the persist-off mode the screen flashes during each event but does not retain the display of the event.

- **F8-See Unknown:** This hot key allows the user to see the unknown scatterer that is being used to generate scattering events. It is only active during the **Guess the Scatterer** mode of the program.

- **F9-New Unknown:** This hot key allows the user to have the program select a new unknown scatterer. It is only active during the **Guess the Scatterer** mode of the program.

- **F10-Menu:** This item activates the menu at the top of the screen.

2.6   *Structure of the Program*

All the source code for this program is contained within the file RUTHERFD.PAS. The program is built around three objects. There is an object named TSCINTIL-LATOR with a data structure and methods to control the display and animation of the scintillator mode. There is an object named THISTOGRAM that has the data structure and methods necessary to display and control a simple rectangular histogram. A third object type, TPOLARHISTOGRAM, is a descendent of the

THISTOGRAM class which has additional data and methods to display and control the polar histogram display on the screen.

The program contains functions for calculating the scattering angle given an impact parameter generated by a random number generator. Some of the types of scattering force (regular n-gon, $1/r^2$ exact, hard sphere, uniform scatterer) use an exact analytic function for the particle trajectories to determine the scattering angle. Other forces ($1/r^2$ integrated, $1/r$, $1/r^3$, Yukawa, and the user defined forces) use an integration of the equations of motion of the beam particle to find the scattering angle.

2.7 Exercises

2.1   **Closest Approach to Nucleus**
The alpha particles used by Rutherford, Geiger, and Marsden to probe matter were 8 MeV alpha particles from ^{214}Po. Consider J. J. Thomson's objection to Rutherford's assumption that the forces between the alpha particle and the atomic constituents remained coulombic at all distances in arriving at his model of the atom. We now know that the forces are not coulombic at distances on the order of the diameter of the nucleus (10^{-15} m). The strong nuclear force dominates at these distances. Calculate the closest approach of an 8 MeV alpha particle to a gold atom nucleus considering only the coulomb force between the alpha particle and the nucleus. Does it come within range of the strong nuclear force?

2.2   **Wave Nature of Alpha Particles**
The radius of the nucleus, as far as the alpha particle is concerned, can be considered to be the classical nearest approach distance (see Exercise 2.1). From elementary optics, we know that the wave nature of light is often unimportant when the scatterer is large compared to the wavelength of light. Similarly, the wave nature of matter is often unimportant when the size of the scatterer is large compared to the wavelength of matter. Calculate the DeBroglie wavelength of the 8 MeV radon alpha particles used by Rutherford, Geiger, and Marsden to probe matter. Do you think that Rutherford's classical derivation of the scattering of the 8 MeV alpha particles is sufficient without the inclusion of wave properties of the alpha particles?

2.3   **Total Scattering Cross Section of Geometrical Scatterers**
The total cross section of the scatterer, σ, is the differential cross section integrated over all solid angle,

$$\sigma = \int d\sigma = \int \frac{d\sigma}{d\Omega} d\Omega. \tag{2.30}$$

Geometrically, this is the area of the incident beam that is scattered in some direction by the scatterer in problems that can be treated with classical physics. Use the program to try to determine the total cross section for the forces under the force menu. The program is two-dimensional rather than three-dimensional, so total cross section and differential cross section are lengths rather than areas. The "unscattered" particles are those that continue on their original path and end up at a zero scattering angle. The ratio of the total cross section width to the total beam width is related to the number of "unscattered" beam particless by

$$\frac{\text{total cross section}}{\text{beam width}} = 1 - \frac{\text{unscattered particles}}{\text{incident particles}}, \quad (2.31)$$

so the total cross section can be estimated from the number of particles that end up at a zero scattering angle, the total number of particles incident, and the width of the beam. Select either the hard sphere or the N-gon scattering force. Generate scattering events for these scatterers and see if the total cross section derived from the scattering results agrees with the expected result that the total cross section is just the projected length of the object in the direction of the beam. For example, the total cross section of the circle should be the diameter of the circle. What condition is imposed on the beam width such that Eq. 2.31 holds true? What condition is imposed on the beam profile such that Eq. 2.31 holds true?

2.4   Total Cross Section for Long Range Forces

Most of the remaining forces in the **Force** menu of the program are long-range forces, forces that are non-zero at arbitrarily large distances from the center of force. Select any of these forces and try to determine the total cross section by the method described in Exercise 2.3. Care must be taken in using the results in this case as the finite resolution of the histogram places small scattering angle results into the zero bin. To determine if this is occurring, reduce the energy of the incident beam of particles. The true unscattered events will remain unscattered even if the energy is reduced. If the number of zero angle scattering events is changed by changing the beam energy, the simulation is probably not determining the true number of unscattered particles. What does your physics intuition tell you about the total cross section of a scatterer that interacts with a long-range force. Does that explain the results?

2.5   Inverse Square Total Cross Section

Integrate the differential cross section derived for an inverse square force in section 2.2 over all solid angles to show that the total cross section for an inverse square force is infinite.

2.6   Uniform Scatterer

Under the **Force** menu there is a **Uniform scatterer** item. Derive the shape of a geometrical surface that would cause this type of scattering.

2.7   Counting Statistics

With any scattering experiment there is a statistical problem to be considered, as should be obvious from running the program. How much data needs to be taken to reach a given uncertainty in the experiment? This is a very important problem in current high energy physics research concerning the top quark[*] and the mass of neutrinos[†] where it takes a great amount

---

[*]Latest top quark results.
[†]Latest massive neutrino experiment.

of time to gather a single event. Select the **Uniform scatterer** from the **Force** menu.

2.8   Electron Scattering

Crowther used beta particles in an experiment similar to Geiger and Marsden's. In the Rutherford experiment, try a beam particle that has the same charge and mass as the electron. For the target, use both a Thomson atom and a Rutherford atom to see if Crowther's experiments were capable of deciding between the two atomic models.

2.9   Geiger-Marsden Experiment

To gain an appreciation of the original experiments performed, if the equipment is not available to you, run the simulation of the Geiger-Marsden experiment using the scintillation detector and count the scintillations as a function of angle. This can be a time-consuming exercise if you are to obtain reasonable statistics on the results with a reasonable angular resolution.

2.10   Numerical Accuracy

The program uses an analytic expression to determine the scattering for the inverse distance potential when it is selected on the input screen. It integrates the force on the particle for user defined potentials. To check the accuracy of the numerical method, run the simulation for a user defined potential that has $1/r$ dependence and compare the values to the analytic results obtained by selecting the $1/r$ potential from the scatterer input screen.

2.11   Hard Sphere Nucleus

Suppose the nucleus were actually a hard sphere of positive charge with a radius of 10^{-13} m. The force on an alpha particle can be modeled by a force of the form

$$F(r) = \frac{1}{r^2} + \delta(R - r). \tag{2.32}$$

Do the 8 MeV alpha particle experiments distinguish between the Rutherford model or this model of the atom? Run a simulation that uses this potential. Compare it to the results of a simulation with pure $1/r^2$ potential. What would you need to change in the experiment that would allow you to distinguish between the two models? In particular, what energy alpha particles would be needed to distinguish between the two nuclear models. Try calculating the necessary energy. Refer to Exercise 2.1. Run a simulation to see if this energy results in significantly different results for the two nuclear models.

2.12   Uniformly Distributed Nucleus

Suppose the nuclear charge were distributed uniformly over a sphere of radius R. Use Gauss' law to determine the electric field as a function of radius from the center of charge. Select **User defined** from the **Force** menu and enter a force of this form into the program. Determine the radius R at which the distribution of the charge over a sphere can be detected with the 8 MeV alpha particles used in Geiger-Marsden experiments.

2.13   Exponentially Distributed Nucleus

Suppose the nuclear charge were exponentially distributed with a decay length of R. Use Gauss' law to show that the electrostatic force on an alpha as a function of distance from the center of charge is given by

$$F(r) = \frac{1 - e^{-r/R}}{r^2}. \qquad (2.33)$$

Select **User defined** from the **Force** menu and enter a force of this form into the program. Determine the decay length R at which this distribution of charge can be detected with the 8 MeV alpha particles used by Geiger and Marsden by entering various values of R and looking for the difference between the results with this potential and Rutherford scattering.

2.14   Failure of the Numerical Method

As discussed in section 2.4, the numerical method used can be fooled with a force function that has arbitrarily large high-order derivatives of the force with respect to position. A good test of this limitation of the numerical method is to test it with a force that has discontinuity. Select **User defined** from the **Force** menu and enter a force of this form into the program. Enter

$$1000 * \mathrm{H}(0.01 - r) \qquad (2.34)$$

into the user defined force input screen. Investigate the scattering distribution for a variety of incident particle energies and beam widths to test whether the numerical method is giving the correct results for the scattering distribution. What might be going wrong?

2.15   Screened Coulomb Potential

Using the user defined force from **Force** menu, enter the function

$$\exp(-r)/r^2. \qquad (2.35)$$

This is the form of the electrostatic force from the electric field of positive point charge surrounded by a exponentially decreasing density of negative charge as the distance from the positive charge increases. The probability distribution of the electron in the hydrogen atom ground state has that dependence. Compare the results of this scattering force to the bare coulomb force. Can the scattering experiment determine the difference between the two scattering forces?

References

1. Hiller, J.R., Johnston, I.D., Styer, D.F. *Quantum Mechanics Simulations, The Consortium for Upper-Level Physics Software*, New York, John Wiley & Sons 1995.

2. Retardation of the α particle from radium in passing through matter. Philosophical Magazine 6; xii, 134-146, 1906.

3. Geiger, H. and Marsden, E., Diffuse Reflection of the α-Particles, Proceedings of the Royal Society, lxxxii: p 495–500, 1909.

4. Thomson, J.J. Cambridge Literary and Philosophical Society, On the scattering of rapidly moving electrified particles. xv: p 465–471, 1910.

5. Crowther, J.A. On the scattering of homogeneous β-rays and the number of electrons in the atom, Proceedings of the Royal Society, lxxxiv: p 226–247, 1910.

6. Rutherford, E., The scattering of α and β particles by matter and the structure of the atom. Philosophical Magazine May 6; xxi:669-688, 1911.

7. Fowles, G.R. *Analytical Mechanics,* 4th edition. New York, Saunders College Publishing, 1986, pp. 141–146.

8. Davis, A.D. *Classical Mechanics.* Orlando, Academic Press Inc., 1986 pp. 174–184.

9. Press, W.H., Flannery, B.P, Teukolsky, S.A., Vetterling, W.T., Numerical Recipes, The Art of Scientific Computing, Cambridge, Cambridge University Press, 1986, p 574.

10. Golub, G.H, Ortega, J.M. *Scientific Computing and Differential Equations: An Introduction to Numerical Methods.* Orlando, Academic Press, 1992, pp. 20–36.

11. Butcher, J.C., *The Numerical Analysis of Ordinary Differential Equations.* New York, John Wiley & Sons, 1987 pp 120–124.

3

Historic Experiments: Electron Diffraction

Michael J. Moloney

3.1 Introduction[1,2,3]

In 1923, Louis deBroglie proposed that electrons might be capable of acting like waves, having a wavelength which depended on their momentum.

   The experiments of Davisson and Germer (in 1926 and 1927) were the first to confirm this wavelike behavior of electrons. The original experiment consisted of bombarding a nickel crystal with a beam of electrons, and observing the scattered electrons at various angles. The work was part of a study begun in 1919 on the elastic scattering of electrons from metals; only around 1926 did it take on the character of looking for diffraction of electron waves.

   An accident broke the vacuum vessel containing the nickel target. In the process of restoring the nickel surface by removing the oxidation from it, the crystal was heated enough to cause recrystallization into large crystallites, instead of the smaller ones which had been there. Anomalies then appeared in the reflected beam.

   To get a clearer picture of the anomalies, a large single crystal of nickel was grown, which was cleaved so that its *(111)* face was exposed. Then Davisson and Germer directed a beam of low-voltage electrons perpendicular to this exposed *(111)* face. They took data on the reflected electrons as a function of angle and found a distinctive peak in their data when the electrons were accelerated through a potential difference of 54 V. This peak was consistent with the crystal structure and the calculated deBroglie wavelength of the electrons.

   G. P. Thomson, also pursuing deBroglie's idea, confirmed wavelike electron behavior in experiments done in 1928. He used cathode rays of energy from 10,000 eV to 60,000 eV passing through thin metal foils. Ring-shaped scattering patterns were observed at small angles in the forward direction. The foils contained many small crystalline regions oriented randomly. The scattering which produced the rings was done by crystalline regions oriented at the correct angles to produce significant constructive interference.

The 1937 Nobel prize was shared by Clinton Davisson and G. P. Thomson for the work they did observing electron diffraction from crystals.

3.2   Electron Waves

In 1923 deBroglie put forward the hypothesis that "particles" might have a wavelength which depends on their linear momentum, p,

$$\lambda = \text{wavelength} = h/p, \tag{3.1}$$

where h = Planck's constant.

Electrons (of charge $-q$) accelerated through a potential difference $-V$ have work qV done on them (q is the unit of electric charge, 1.61×10^{-19} C). This work changes their kinetic energy, and if the electrons start from rest, they finish with a kinetic energy of

$$\text{kinetic energy} = \frac{p^2}{2m} = qV, \tag{3.2}$$

where m is the mass of the electron. Thus, the electron's wavelength, according to deBroglie's hypothesis should be

$$\text{wavelength} = \lambda = \frac{h}{(2mqV)^{1/2}}. \tag{3.3}$$

3.3   The Relativistic Correction to Electron Wavelength

The electron's kinetic energy changes due to the work qV done on it, as it is accelerated through a potential difference V. When qV is no longer negligible with respect to the electron's rest mass energy mc^2 (511 keV), relativity effects become important, and a correction to the electron's momentum and wavelength must be made. After acceleration from rest, the electron's kinetic energy equals qV, where q is the magnitude of the electron charge. The electron's total energy E is the sum of its rest mass energy mc^2 (c is the speed of light in vacuum) and its kinetic energy:

$$E = qV + (mc^2). \tag{3.4}$$

The electron's total energy E is related to its linear momentum p by the following equation:

$$E^2 = (pc)^2 + (mc^2)^2. \tag{3.5}$$

When E is eliminated between Eqs. 3.4 and 3.5, we find a revised equation for p:

$$p = \frac{((qV)^2 + 2qVmc^2)^{1/2}}{c}. \tag{3.6}$$

When the accelerating voltage is not small compared to 511 kV, Eq. 3.6 must be used instead of Eq. 3.2 to determine the momentum from the accelerating voltage V, so the wavelength can then be determined from Eq. 3.1.

3.4 FCC Lattices and Planes

The nickel crystal used by Davisson and Germer is "face-centered cubic" (fcc), and many of the metal foils used by G. P. Thomson are also fcc lattices. The fcc structure has atoms at the 8 corners of a cube and there is an atom in the center of each of the 6 faces. The side length of the cube is known as the *lattice constant,* "a." Crystal planes are denoted by (*hkl*), where *h, k,* and *l* are integers derived from the reciprocals of the intercepts of the plane on the three axes. Equal intercepts on all three axes give the (111) plane. A plane which intersects only one axis is denoted by (*h*00), and one which intersects only two axes by (*hk*0). The distance between (*hkl*) crystal planes for cubic systems is given by[6]

$$d_{hkl} = \frac{a}{(h^2 + k^2 + l^2)^{1/2}}. \tag{3.7}$$

This chapter's view of fcc lattices is two-dimensional only. In the (*hk*0) scattering version, the lattice is square, with an atom-to-atom separation half the lattice constant. For (111) scattering, the geometry is as shown in Figure 3.1. Distance AD equals the lattice constant, a. Six atoms are shown in the (111) plane, along lines parallel to CD. When these lines are viewed end-on, they make up a column whose spacing is $a(\frac{3}{8})^{1/2}$. Length AB is the distance between successive (111) planes. AB equals $\frac{a}{3^{1/2}}$, and is the spacing between columns of (111) atoms, or it may be thought of as the separation of atoms in a row.

The face-centered cubic nickel crystal has a lattice constant of 3.51 Å. The (111) face has lines of atoms parallel to one another spaced 2.15 Å apart. These

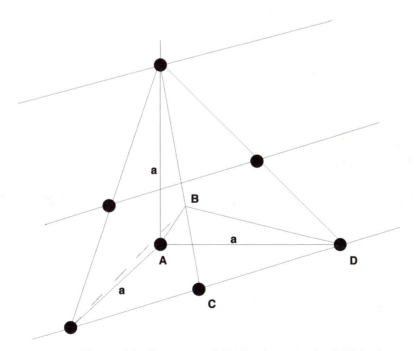

Figure 3.1: Geometry of (111) planes in the FCC lattice.

lines of atoms act like a reflection grating of spacing 2.15 Å, giving the distinctive lobe of reflected electrons at an angle of 130° to the incoming electron beam direction when the electron kinetic energy is 54 eV.

Grating spacings of the various metal foils used by G. P Thomson were in the vicinity of 4 Å. The fcc lattice of aluminum (lattice constant = 4.05 Å) is used to simulate G. P. Thomson's scattering experiments. Figure 3.2 illustrates scattering from small crystallites of aluminum. The rings indicate constructive interference of scattered electrons from randomly oriented crystallites of aluminum, either in powdered form or in a foil.

3.5   Scattering From Rectangular Arrays

This section considers scattering from a rectangular array of atoms where the distance between columns is d_1 and the distance between rows is d_2. The scattering simulation is restricted to two dimensions. For fcc lattices, it accurately models scattering from $(hk0)$ planes, and Bragg scattering from the (111) planes. When a beam comes in perpendicular to the (111) plane, however, it encounters staggered rows of atoms in successive (111) planes. The treatment which follows does not account for staggered rows.

3.5.1   Definitions

All angles are referred to the direction of the incoming beam:
   ϕ is the direction of a crystal plane from the incoming direction.
   θ is the scattering angle from the incoming beam direction.
   The distance between rows of atoms in a single plane is d_1.
   The distance between successive planes is d_2.

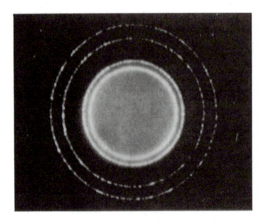
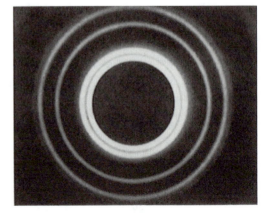

Figure 3.2: Diffraction pattern from powdered aluminum (after Fig. 44-4, p. 1159, ref. 5). **Left**: X-rays. **Right**: Electrons.

3.5.2 Basic Path Difference Relationships

Scattering from two atoms A and B in a crystal plane is considered in Figure 3.3. The incoming and outgoing beams are parallel to each other, and the path difference between the two beams is seen to be

$$\text{path difference}_{AB} = d_1\left(\cos(\theta - \phi) - \cos\phi\right). \tag{3.8}$$

A more useful version of this expression for the path difference can be obtained with a couple of trigonometric operations, and gives

$$\text{path difference}_{AB} = 2d_1 \sin\frac{\theta}{2}\sin\left(\frac{\theta}{2} - \phi\right). \tag{3.9}$$

Atoms along a line perpendicular to d_1 lie in the next crystal plane a distance d_2 away from the AB line (shown in Fig. 3.4). The line from A to the next crystal plane is perpendicular to the AB direction and makes an angle $\phi + \pi/2$ with the incoming beam. Equation 3.9 is easily modified to take account of this new angle and give the path difference between beams which scatter from successive planes:

$$\text{path difference}_{\perp AB} = 2d_2 \sin\frac{\theta}{2}\cos\left(\frac{\theta}{2} - \phi\right). \tag{3.10}$$

Equation 3.9 may be thought of as path difference due to adjacent atoms in the same row, and Eq. 3.10 is then the path difference due to adjacent atoms in the same column, as illustrated in Figure 3.4.

Each wavelength of path difference results in 2π of phase difference between the beams. Calling the phase difference δ, we may write

$$\delta = \text{phase difference} = \frac{2\pi}{\lambda}\,(\text{path difference}). \tag{3.11}$$

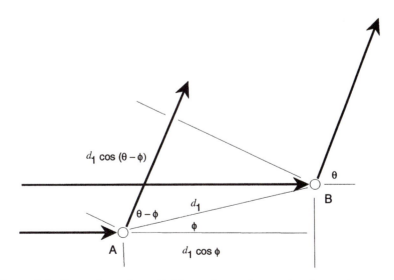

Figure 3.3: Scattering of parallel incoming and outgoing beams by two atoms.

3.5.3   Conditions for Intense Scattering

Parallel incoming beams of x-rays or electrons scattering from a crystal will produce strong constructive interference as parallel outgoing beams if every scattered beam is in step (or "in phase") with all the others.

Equation 3.9 gives the phase difference between beams scattered from adjacent rows of atoms in a plane, and shows that if $\theta/2$ equals ϕ, the phase difference will be zero for all beams scattered from rows in that plane. This corresponds to Figure 3.4, where the scattered (outgoing) beam makes the same angle, ϕ, with the crystal plane as does the incoming beam, so that the beam appears to have had a mirror-like reflection from that plane.

But the beams scattered from the first plane must be in phase with the beams scattered from the second and successive planes in order that all scattered beams be exactly in phase. Equation 3.10 shows the extra path between two adjacent planes. When this extra path is an integer number of wavelengths, and when mirror-like scattering occurs, the conditions are met for intense scattering. This circumstance is known as "Bragg reflection" or "Bragg scattering":

$$2d_2 \sin \phi = n\lambda, \tag{3.12}$$

where n is an integer greater than zero, and ϕ is shown in Figure 3.4.

Figure 3.5 shows the situation corresponding to the original version of the Davisson-Germer experiment, where a beam comes in perpendicular to a single plane of atoms (the exposed (111) plane of a nickel crystal), and the scattered beams travel to the distant detector parallel to one another. Rows of atoms are separated by d_1 and the situation is very similar to transmission through a grating. The path difference between scattered beams from adjacent rows is $d_1 \sin \alpha$. The scattered beams will all be in phase if this path difference is an integer number of wavelengths λ:

$$d_1 \sin \alpha = n\lambda. \tag{3.13}$$

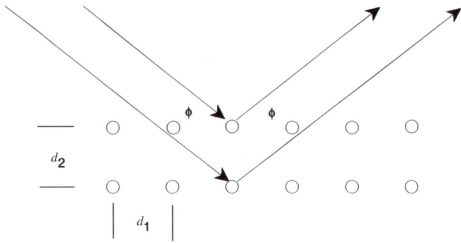

Figure 3.4: Bragg (mirror-like) scattering from successive crystal planes.

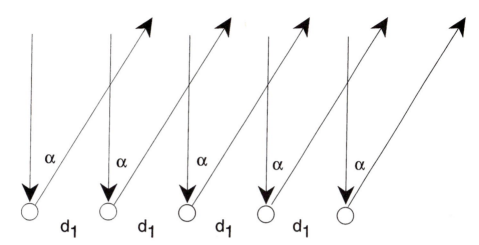

Figure 3.5: Scattering from a single crystal plane at normal incidence (behavior like a diffraction grating).

This condition disregards the path difference between beams scattering from successive planes parallel to the face of the crystal separated by distance d_2: it deals only with the surface layer. The electrons in the Davisson-Germer experiment were of very low energy and could have been expected to penetrate very little into the nickel crystal.

3.5.4 Destructive Interference; Sharpness of Peaks

The scattered wave amplitude from N equally spaced elements in a line depends on the phase difference δ (see Eqs. 3.9 and 3.10) between adjacent atoms (or rows, or planes, in the 2-D simulation).

The N scattered beams can be combined by adding their "phasors," much like vectors, head to tail. If the beams are all in phase, the phasors add in a straight line; if not, then the angle between phasors is just the phase difference δ between them. The N phasors can add up to zero if the head-to-tail train wraps around into a complete circle (really a polygon of N sides). This gives total destructive interference of all N scattered beams. This destructive polygon results from the total phase difference between the first and Nth scattered beam of 2π:

$$N\delta = 2\pi. \tag{3.14}$$

If this equation is divided by 2, it expresses the fact that total destructive interference can result from a phase difference between the first and the middle scattered beam of π, causing these beams to cancel out. That would lead to complete cancellation of the beams in the first half, with those in the second half, and there would be no amplitude at all when all scattered beams were added together.

The exact mathematical expression for the N-beam sum with a phase difference δ between each component is given[4] by

$$\text{total amplitude due to } N \text{ atoms} = A\frac{\sin(N\delta/2)}{\sin(\delta/2)}, \tag{3.15}$$

where A is the scattered amplitude due to one atom by itself. The phase with respect to that of the beam scattered from A is given by

$$\text{total phase due to scattering by } N \text{ atoms in a line} = (N-1)\delta/2, \tag{3.16}$$

where δ is the phase difference between beams scattered from adjacent atoms in the line.

As the scattering angle θ changes, the "Bragg" scattering maximum will gradually die away to zero, when total destructive interference sets in (Eq. 3.14). The δ causing total destructive interference could be due to row-to-row path difference (Eq. 3.9), or it could be due to successive-plane path difference (Eq. 3.10). Either way, this depends on the number, N, of successive beams to be added. One of the suggested exercises will follow up on this idea.

3.5.5   Scattering Due to All Atoms in One Crystal

To obtain the scattering from all atoms in a crystal, first obtain the amplitude A_{row} due to scattering from all atoms in one row, imagining a "reference" atom A which is at the start of the row:

$$A_{\text{row}} = A_{\text{atom}} \sin(N\delta(r)/2)/\sin(\delta(r)/2), \tag{3.17}$$

where $\delta(r)$ is obtained from angle $\phi(r)$ which a row makes with the x-axis. The phase of this scattering amplitude with respect to the beam scattered from A is

$$\chi(r) = (N_{\text{row}} - 1)\delta(r)/2, \tag{3.18}$$

where N_{row} is the number of atoms in one row. The amplitude from the entire crystal is obtained by adding amplitudes from each row together across the columns. Recall that each of these amplitudes has magnitude $A(\text{row})$ and has a phase difference of $\chi(r)$ with respect to the beam scattered from A.

The "column" phase difference $\delta(c)$ for scattered beams from atoms in adjacent rows is based on the angle $\phi(c)$ which each column makes with the x-axis. The phase difference $\chi(c)$ is obtained from the phase difference $\delta(c)$ just as in Eq. 3.18,

$$\chi(c) = (N_{\text{col}} - 1)\delta(c)/2, \tag{3.19}$$

where N_{col} is the number of atoms in a column. The amplitude from the entire crystal is

$$A_{\text{crystal}} = A_{\text{row}} \sin(N\delta(c)/2)/\sin(\delta(c)/2), \tag{3.20}$$

or substituting for A_{row},

$$A_{\text{crystal}} = A_{\text{atom}} \left[\frac{\sin(N\delta(c)/2))}{\sin(\delta(c)/2)} \frac{\sin(N\delta(r)/2)}{\sin(\delta(r)/2)} \right]. \tag{3.21}$$

The phase of A_{crystal} with respect to A_{row} is $\chi(c)$. And the phase of A_{row} with respect to A_{atom} is $\chi(r)$. Thus, the phase of the total amplitude scattered from the entire crystal is

$$\chi(c) + \chi(r) = (N_{\text{row}} - 1)\delta(r)/2 + (N_{\text{col}} - 1)\delta(c)/2 \tag{3.22}$$

with respect to the beam scattered from A (i.e., with respect to A_{atom}).

3.5.6 Scattering From All Crystals

Each crystal has a reference atom which is at (x, y) with respect to some overall origin (0,0). Thus, there is a phase difference between the reference atom of each crystal and a hypothetical reference beam scattered from the overall origin.

Each different crystal has an amplitude given by Eq. 3.21, and an overall phase with respect to the hypothetical beam:

$$\gamma = \text{overall phase} = \text{``}xy\text{'' phase} + \chi(c) + \chi(r). \tag{3.23}$$

The scattered waves from various crystals are combined by taking account of their phases

$$A_x(\text{total}) = \sum_i [A_{\text{crystal},i} \cos \gamma_i], \tag{3.24}$$

$$A_y(\text{total}) = \sum_i [A_{\text{crystal},i} \sin \gamma_i], \tag{3.25}$$

$$A(\text{total}) = [A_x(\text{total})^2 + A_y(\text{total})^2]^{\frac{1}{2}}. \tag{3.26}$$

This is the overall scattered amplitude from all crystals.

3.6   Simulation of Davisson-Germer Scattering

A two-dimensional "crystal" with 8 rows and a variable number of columns of atoms is used in the simulation. The rows are separated by $d_1 = 2.15$ Å, the distance in a nickel crystal between rows of atoms within the (111) plane, and the columns are separated by $d_2 = 2.027$ Å, the distance between successive (111) planes.

The electron accelerating voltage can be adjusted over a range of 0 to 100 V. This allows the user to observe scattering in the same general way as was done by Davisson and Germer.

The Davisson-Germer experiment was a reflection experiment, so that forward transmission angles do not apply. Forward scattering angles (near zero) have been left on the sliders for the general interest of the user, and to keep the same angles as used for the G.P. Thomson scattering simulation. The closest approximation to the actual Davisson-Germer experiment is a single column of atoms, corresponding to the exposed (111) plane of the crystal.

Most textbooks refer to the $\alpha = 50°$ scattering at which Davisson and Germer obtained their conclusive results with electron kinetic energies of 54 eV; this angle corresponds to $\theta = 130°$ of scattering from the direction of the incoming electron beam.

A variation of the Davisson-Germer experiment was done with the beam striking the crystal face at an oblique angle. When the accelerating voltage was scanned, intense maxima occurred at an angle corresponding to Bragg reflection from the (111) planes. The **VScan** hot key provides the variation in voltage for this experiment, and is the subject of a suggested exercise. In this case, it is desirable to use more than one column of atoms, to simulate scattering from successive (111) planes in the crystal.

3.7   G. P. Thomson Scattering

G. P. Thomson sent a beam of high energy (on the order of 60,000 eV) cathode rays through thin metal foils, and observed forward scattering as concentric rings on a photographic plate. Because of the short wavelength, these produced intense scattered beams at small angles in the forward direction. Typical thicknesses of Thomson's foils were several hundred angstroms. This simulation uses square two-dimensional crystallites oriented randomly to the direction of the incoming electron beam.

The data of G. P. Thomson is explained by the same reasoning given for "Bragg" scattering of x-rays from atomic planes. Mirror-like reflections from each plane insure that all atoms in the same plane give scattered beams which are in phase, because the extra path between beams reflecting from that plane is zero.

G. P. Thomson's beams of electrons all had the same energy and wavelength. They encountered small crystallites at all different angles ϕ to the beam. The crystallites have many possible d_2 values between the different crystal planes, and different angles ϕ will satisfy the extra-path condition for different d_2 values, which gives rise to the bright rings seen by G. P. Thomson (illustrated in Fig. 3.2).

[The rings in Figure 3.2 represent scattering from the following planes, in order of increasing ring radius: (111), (200), (220), and (311).]

The program provides "aluminum" crystallites oriented to simulate scattering from ($hk0$) planes: an $N \times N$ square array 2.025 Å (half of the "lattice constant") on a side. The user may select N atoms from 2 to 100 within a crystallite, and the number of crystallites may be varied from 1 to 100. This is a realistic range, since Thomson used foils which were about 400 Å thick.

3.8 X-Ray Diffractometer

In an x-ray diffractometer, x-rays undergo Bragg reflections from a crystal in the same way that electrons undergo Bragg reflections from the crystallites in a metal foil. Thus, to show the same physics in a different setting, the program includes an x-ray diffractometer simulation.

X-rays are generated in a diffractometer by accelerating electrons through a potential difference $-V$ and having them strike a target (often copper). X-rays are generated in the collision process, up to a maximum energy of qV.

The scattering crystal is mounted and driven so its angle of inclination to the incoming beam is always half of the scattering angle to the detector. That is, the detector rotates through twice as much angle as the scattering crystal. This means that crystal planes parallel to the face of the mounted scattering crystal always behave so the scattered x-rays heading toward the detector have made a mirror-like reflection.

The x-rays are collimated and scatter from the crystal to the detector as shown in Figure 3.6. The same set of crystal planes is always presented to the x-ray beam, but as the angle θ changes, the path difference between beams scattering from successive planes also changes. Only when this extra path is an integer number of wavelengths is there constructive interference between the beams going to the detector.

The "crystal" used in the simulation is 4×8 atoms, with the same spacing as the fcc structure of lithium fluoride (LiF). Scattering from this crystal can be seen in the ($hk0$) planes.

It is interesting that the x-rays act like waves in scattering from the crystal, and act like particles when they enter the detector about 1 m away from the crystal.

3.9 Details of the Program

a. **"Instrument" parameters**
These operate much like the controls of an actual instrument. Both coarse and fine accelerating voltage adjustments are available, so that precise wavelengths can be selected. The angular range of the detector is adjustable, and the detector gain is adjustable. Detector gain settings carry over from one type of run to another, and may need to be reset when going between different types of runs.

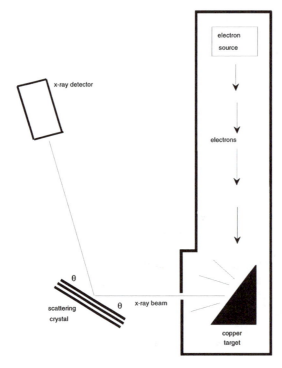

Figure 3.6: X-ray diffractometer.

b. **"Target" Parameters**

In Davisson-Germer scattering, only target orientation ("Target Direction") is adjustable. In Thomson scattering, targets (crystallites) are randomly placed in x and y, and randomly oriented in angle. The range of possible angles is controlled by the user, from 1° to 90°. This permits the user to keep the crystallites oriented close to the forward direction, so that small angle forward scattering, as seen by Thomson, can be studied in some detail. Crystallites are placed randomly in a region 0.1 radians wide centered on the "central angle". This shows how randomly oriented crystals near the Bragg scattering angle affect the shape of the scattering peak.

The number of crystallites can be varied from 1 to 100. The number of atoms per row or column of a crystallite can be varied from 2 to 100. This lets the user get a feel for the effect of crystal size on the diffraction pattern (mainly on the "sharpness" of the scattering peaks). This is to help understand how Thomson's well-defined rings were formed.

The user may study a given group of crystallites again and again, by varying gain, voltage, angular scan limits, etc. Then a new group of crystallites can be selected via the hot key **NewXtals**. This feature is intended to let the user examine the reproducibility of the results between groups of crystallites.

The user can also establish a single crystallite in a given orientation. When only one crystallite is selected, the "center angle" of the crystallites will be the angle at which the single crystallite is oriented.

c. **Screen Hot Keys**

AngScan initiates a scan of the detector over a selected range of angles. In Davisson-Germer scattering, **VScan** initiates a scan of voltage from the selected value to 16 times the selected value, with the detector held at 90° to the incoming beam. The detector output is graphed versus the square root of the voltage. All scans are carried out in exactly 340 increments. Angle markers 0.4° wide are laid down every degree, 0.2° to either side of center. Positions at zero and 90° are denoted in black, multiples of 30° are denoted in white. Multiples of 5° and 10° are set off with slightly different heights. It is expected that the user could determine angle in a scan to around 0.2°.

For G. P. Thomson scattering, the **Planes** hot key lets the user select either the (111) plane spacings or the (*hk*0) plane spacings. For Davisson-Germer scattering, the **Nr Planes** hot key lets the user have only one plane of atoms in the nickel crystal, or four planes. A single plane is the closest approximation to scattering from the surface only, without interference from deeper planes in the crystal. It behaves like a diffraction grating, and produces a scattering pattern symmetric about 90°. One should definitely use four planes for the **VScan** work, where Bragg scattering is going on.

3.10 Exercises

3.1 **Repeat the Davisson-Germer Experiment**

a. Repeat the Davisson-Germer experiment as reported in the literature. The display initially indicates only a single (111) plane will be scattering the electrons, and the target crystal is at 0°. Keep these settings and do scans at 40, 44, 48, 54, 60, 64, and 68 V. Record both the wavelength in Å, and the relative height of the peak at 130° for each voltage.

b. Treating the spacing of 2.15 Å as a grating spacing, calculate the extra path in Å at a scattering angle of 130°. (Keep in mind that this is backward scattering from rows of atoms on the (111) surface, and not Bragg scattering from the entire crystal.)

c. Write down the ratio of extra path to wavelength for each of the voltages in part a. (You should obtain a nearly integer relationship at 54 V.)

3.2 **A Variation of the Davisson-Germer Experiment**

a. Set up for Davisson-Germer scattering. Rotate the target to an angle of 60°. Draw a sketch to show that Bragg reflections should be seen (if the wavelength is right) at a scattering angle of 120°. (Hint: Bragg reflections involve mirror-like reflections from each crystal plane.)

b. Vary the voltage (try increasing it by about 1 V at a time) and do a series of angle scans until you have a very large scattering peak at 120°. From the fact that the electron beam makes a 30° angle with the crystal planes, and given that the distance between crystal planes is 2.15 Å, work out the extra path in angstroms taken by a beam of electrons between crystal planes. Compare this extra path to the wavelength you are using.

c. Predict another voltage which will also give a large scattering peak at 120°. Record your prediction and try it out on the simulation. If a large peak fails to appear with your predicted voltage, check the wavelength, to be sure it is correct. Then try again until you achieve the desired results. Record your predictions and results.

3.3   Scattering from Several (111) Planes

This exercise explores changes that would occur in the Davisson-Germer experiment due to scattering from several (111) planes, rather than just the surface (111) plane.

a. Select Davisson-Germer scattering, leave the target angle at 0°, use the **Nr Planes** hot key to select four (111) planes (instead of just one), set the accelerating voltage to 54 V, and do an angle scan. Write down the angle at which the large scattering peak occurs.

b. Why does the peak occur at essentially the same angle as for a single crystal plane (as you did in the first exercise), now that four planes are involved? To answer this, you must explore the phase difference between beams scattered from different planes. Why is the Bragg scattering formula *not* appropriate? Write down your reasoning, along with a sketch of the scattering situation with four (111) planes involved.

c. Determine the correct relationship to use, so that you can calculate the correct scattered amplitude from four planes. Demonstrate quantitatively why the scattered amplitude will be very nearly a maximum with four planes. (Hint: Phase difference is the key here, as in many scattering situations.)

3.4   Qualitative Explanation of Scattering at $\theta = 0$

a. Sketch a rectangular crystal, showing one line of its atoms making an angle ϕ with an incoming beam. Sketch the outgoing beam scattering in the $\theta = 0$ direction. Accompany your sketch with an explanation of why there will always be intense scattering at $\theta = 0$ no matter what the angle ϕ. (Don't use any equations.)

b. Support your discussion in part a by examining Eqs. 3.9 and 3.10.

3.5   Bragg Scattering From the (200) Plane of Aluminum

a. Using a lattice constant a for aluminum of 4.05Å, calculate the distance between (200) planes d_{200} from Eq. 3.7.

b. Using $\lambda = 0.497$ Å (the default value in G. P. Thomson mode), calculate the angle ϕ for Bragg reflection in first order ($n = 1$).

c. Select **G. P. Thomson** under **Run**. Select a single crystallite and then use the **Center Angle** slider to set it to the angle ϕ from part b.

d. Do an angle scan and see whether you have intense scattering at angle 2ϕ. Repeat a–c if necessary. Record and turn in your work when the angle scan shows intense scattering (essentially full height) at angle 2ϕ.

3.6   Bragg Scattering From the (111) Plane of Aluminum

Select **G. P. Thomson** mode, set the voltage to 600 V, and use a single crystal. Use the **Planes** hot key to set the crystal parameters for (111) plane scattering. [You will see (111) within the crystal, instead of ($hk0$).] Calculate the correct angle ϕ for the crystal to produce first-order Bragg scattering with the (111) plane of aluminum, and set this angle using the **Center Angle** slider. Write down the angle ϕ. Run an angle scan to verify that a large Bragg peak occurs at 2ϕ.

3.7   Determining the Plane of ($hk0$) Bragg Scattering

Use the default **G. P. Thomson** parameters and determine from the angular scan data and other information on the screen what plane of ($hk0$) scattering is responsible for the large Bragg peak in addition to the one in the center. Draw a diagram showing the crystal oriented at 55° to the incoming beam, and clearly identify how the scattering occurs. Draw incoming and outgoing rays for the electron beam, and show how the Bragg condition is met.

3.8   Derive Eq. 3.12

a.   Base the derivation entirely on Figure 3.4. Make a sketch, do the necessary geometry, and explain why Eq. 3.12 must be satisfied.

b.   Base the derivation on Eqs. 3.8 and 3.9, using mathematical ideas rather than geometrical ones.

3.9   Measured Bragg Scattering From Photographs

a.   If the scattering foil is a distance L from the plane of the photo (for example, Fig. 3.2), show that the radius R of a ring due to crystal planes separated by distance d is given by $2L \tan(\sin^{-1}(\lambda/2d))$. (This is for first-order scattering; the rings represent strong constructive interference.)

b.   Show that if the angles are all small for rings in the photograph, each ring radius will be inversely proportional to the plane spacing d.

3.10   Measured Bragg Scattering From Figure 3.2 (left)

a.   With a lattice constant of 4.05 Å, use Eq. 3.7 and calculate d_{hkl} values for the (111), (200), (220), and (311) planes of aluminum. These four planes have the largest separation values of any planes which give constructive interference rings in all fcc lattices. (Destructive interference automatically occurs in other fcc planes with larger spacings than these four.)

b.   It is claimed that Figure 3.2 (left) was made using a wavelength of 0.709 Å. Based on this wavelength, calculate the angles θ at which large Bragg reflections will occur for each of the planes in part a.

c.   Measure the diameters of all four circular patterns in Figure 3.2 (left) as carefully as possible, and record them.

d.   Assume the rings increase from smallest to largest in the order of planes given in part a, and that the film was a distance of 0.100 m from the aluminum. Calculate the radius of each of the four rings. (The distance of 0.100 m is not necessarily correct, but your radii should still be in the right ratios to one another.)

e.   Take r_1 to be the smallest radius and make a table of ratios r_4/r_1, r_3/r_1, and r_2/r_1, from part d and from the previous exercise.

f.   If the claim that $\lambda = 0.709$ Å is not correct, is the true wavelength larger or smaller?

3.11   Measured Bragg Scattering From Figure 3.2 (right)

a.   It is claimed that Figure 3.2 (right) was made using electrons accelerated through a potential difference of 600 V.

b.   Determine whether the evidence in Figure 3.2 (right) is consistent with this claim, following the same general procedure as in the previous exercise.

c.   If an accelerating voltage of 600 V is not correct, was the true voltage larger or smaller?

3.12   Large Peaks in the Voltage Scan

In Davisson-Germer scattering, use the default voltage of 40 V, and set the target angle to 45°. Then do a voltage scan. Record the voltage scan pattern by sketching on a piece of paper where the peaks and valleys are located.

a.   Work out and write down the path difference in angstroms between electron waves which strike the first plane of the crystal, and those which strike the second plane of the crystal.

b.   For two adjacent large peaks, determine the approximate voltage by reading from the "strip chart" scale as well as you are able. Work out the wavelength for each voltage. Determine a simple relationship between the two wavelengths.

c.   Explain how the wavelengths in part b produce the large peaks. Include a sketch of the crystal planes (parallel to the sides of the rectangle). Show the incoming and outgoing beams. Clearly identify the path difference between the two beams which reflect in a mirror-like fashion from the crystal planes.

d.   Demonstrate that Bragg scattering from the 2.027 Å distance between (111) planes is causing these peaks.

e.   Explain how the wavelengths in part b produce the large peaks.

3.13   Relativistic Wavelength Correction

a.   Set up for G.P. Thomson scattering, with an accelerating voltage of about 15,000 V. Record the wavelength given on the screen. Calculate the wavelength using Eq. 3.3. Indicate the percent difference between the wavelength you calculated and the one on the screen. Now use Eq. 3.6 for the momentum, calculate the wavelength from Eq. 3.1, and write down the percent difference between your result and the wavelength on the screen.

b.   Repeat part a with 100 V as the accelerating potential difference.

c.   What do you conclude about how the wavelength on the screen is calculated?

3.14   **Phase Difference at Small Angles**

This exercise considers scattering for small angles θ and ϕ (less than 20° or 25°). Under these conditions, take $\sin x = x$ and $\cos x = 1$.

a.   Show that the phase difference δ due to row-to-row interference of N beams is approximately

$$\delta = 2\pi N \frac{d_1}{\lambda} \theta(\theta/2 - \phi). \qquad (3.27)$$

b.   Show that the phase difference δ due to plane-to-plane interference of N beams is approximately

$$\delta = 2\pi N \frac{d_2}{\lambda} \theta. \qquad (3.28)$$

c.   If Bragg scattering occurs at θ_o, show that we must have

$$\theta_o = 2\phi, \text{ and } d_2\theta_o = n\lambda, \qquad (3.29)$$

where n = 1,2,...

d.   Will the phase difference from part a or part b first extinguish the scattered beam near θ_o? Support your answer with quantitative reasoning.

3.15   **Small-Angle Bragg Peak Widths**

The "peak width" is the difference in angles closest to a Bragg peak where the intensity is zero.

a.   Use the results of the previous exercise to show that N times the angular width of the peak (between the intensity zeros on either side of θ_o) approximately equals θ_o, for first-order ($n = 1$) peaks.

b.   Work out the relation between peak width and θ_o for any order of Bragg scattering (any integer value of N) at small angles.

3.16   **Bragg Peak Widths Near $\theta = 0$**

a.   Show that there will be a Bragg peak at $\theta = 0$ for any orientation angle ϕ of the crystal.

b.   Find a relation between d_1, d_2, N, and λ for the width of the "central" Bragg peak at $\theta = 0$.

3.17   **Measured Sharpness of Bragg Peaks**

a.   Select G. P. Thomson scattering with a single crystallite, and set its angle to 10°. For (200)-plane scattering, calculate the wavelength needed so that first-order (n = 1) Bragg reflection will occur. Set the accelerating voltage to give the desired wavelength. Record your work.

b.   Make sure the **No. in a row or col (N)** slider indicates that N = 6. Run an angle scan. After you have run one scan, set the scan limits so that you get all of the scattering peak, and a little bit beyond. Do another run, and then record the angle of the peak, and both angles where the scattering intensity is zero.

c.   Use the slider to set $N = 20$. Redo an angle scan, and record the peak value, and both angles where the scattering intensity vanishes.

d.   Repeat part c by setting $N = 40$ on the slider, and recording the results.

e.   It was claimed when discussing N atoms in a line that the sharpness of the peak increases with increasing N. Calculate the product of N and the width of the peak (angular distance between the zeros of intensity) for parts b–d.

f.   Calculate the numerical value of $\delta(r)$ for row-to-row interference for the conditions of part c (when $N = 20$).

g.   Calculate the numerical value of $\delta(c)$ for plane-to-plane interference for the conditions of part c (when $N = 20$).

h.   Discuss the results of part e and f. Which factor caused the peak to be extinguished? Cite all the evidence in favor of your conclusion.

3.18   X-ray diffractometer Valleys

a.   Set up the x-ray diffractometer and do an angle scan. You should observe a series of regular peaks from zero to $180°$. Scan from zero to about $90°$ to see the peaks "spread out" somewhat. Record the wavelength used, and sketch the structure of the peaks and valleys on a piece of paper. Change the wavelength by at least a factor of 2 and do another angle scan. Does the number of valleys between large peaks change when you change the wavelength?

b.   In order to explain peaks and valleys on the basis of interference between the four planes of atoms, and the eight rows per plane, record the wavelength you are using, the x-ray energy, and the angles of the valleys in between large peaks. Work out the extra path the x-rays travel between rows of atoms at two of the angles you have recorded.

c.   For each of the extra path values in part b convert the extra path into a phase difference. Carefully explain this phase difference. Is it the phase difference between x-rays from the first and last rows in a plane? If not, exactly what is it?

d.   Construct an argument to show that this phase difference will account for the zero value of intensity at the places you have recorded. Include a sketch with this argument.

3.19   X-Ray Diffractometer Peaks

a.   Repeat the previous exercise, scanning from $0°$ to $90°$. Adjust the voltage and wavelength until exactly four peaks occur between $0°$ and $90°$ (one is at $0°$, one is at $90°$, and two are in between $0°$ and $90°$). Record the wavelength, and the angles where the intensity is a maximum.

b.   Calculate the extra path between planes and also between rows within a plane for each angle where a peak occurs.

c.   Explain why these path differences cause a maximum of intensity. Include a simple sketch at one angle showing incoming and outgoing beams, to support your discussion.

References

1. Semat, H., Albright, J.R. *Introduction to Atomic and Nuclear Physics.* New York: Holt, Rinehart, and Winston, 1972, pp. 155–160.

2. Richtmyer, F.K., Kennard, E.H., Lauritsen, T. *Introduction to Modern Physics.* New York: McGraw-Hill, 1955, pp. 180–184.

3. Krane, K. *Modern Physics.* New York: John Wiley & Sons, 1983, p. 91.

4. Marion, J.B., Hornyak, W.F. *Physics for Science and Engineering.* Philadelphia: Saunders, 1982, pp. 1147–1148.

5. Halliday, D., Resnick, R., Walker, J. *Fundamentals of Physics*, extended version, 4th edition. New York: John Wiley & Sons, 1993.

6. Cullity, B.D. *Elements of X-Ray Diffraction.* Reading, MA: Addison-Wesley, 1956.

4

Special Relativity

Douglas E. Brandt

4.1 Introduction

A common topic covered in modern physics courses is special relativity. This topic is quite often introduced in introductory physics courses, but is studied in greater depth and with greater mathematical sophistication in modern physics courses.

The topic of special relativity encompasses nothing more than the rules used to relate measurements of physical quantities of interest made by observers in relative motion in the absence of gravitational fields. In your first introductory physics course you probably learned within the first few weeks that motion is relative. Two observers moving with respect to each other measure different velocities for objects that are observed by both. A person inside a moving car holding a ball would say the ball is at rest. A person standing on the ground outside the car would say the ball moves at the same velocity as the car. The mathematics that you learned to relate the measurements of one observer to the those of the other are called Galilean transformations.

The Galilean transformation rules appear to be the correct transformation rules of nature to us because they agree with our everyday experiences. Our everyday experiences are usually limited to relative motions of objects and observers that are quite small compared to the speed of light. However, when considering light, the rules of electromagnetism, and objects with relative speeds near the speed of light, some conflicts arise if Galilean transformations are used to relate the measurements made by observers in relative motion. Probably the one most readily understandable phenomena demonstrating the problem is that all observers always measure the same speed for light. Galilean transformation rules always result in different velocity measures for observers in relative motion. If observer A measures the speed of light to be 3×10^8 m/s and observer B moves at 1×10^8 m/s in the direction that the light travels relative to observer A, Galilean transformation rules give 2×10^8 m/s as the speed of light relative to observer B. The speed of light is always measured to be the same by all observers, so there must be a problem with how Galilean transformations relate velocities of observers in relative motion.

A transformation rule relating the measurements made by different observers in relative motion that predicts that the speed of light will be measured to be the same by all observers is called the Lorentz transformation. The Lorentz transformation also approximates the Galilean transformation very precisely at relative speeds we encounter in everyday life. There have been no measurements yet made that are inconsistent with Lorentz transformations, so physicists currently believe that Lorentz transformations are the correct rules for relating measurements made by different inertial observers when gravitational fields are negligible.

As we have ingrained in our mind the Galilean-like behavior of our everyday experiences with moving objects, it is difficult to accept some of the non-Galilean results from the application of the Lorentz transformation. Introductory physics courses usually introduce students to Lorentz length contraction, Lorentz time dilation, Lorentz velocity transformation, and the transformation rules for energy and momentum. Each of these may be confusing or even troubling because of our experiences in which Galilean transformations appear to be correct.

4.2 *Inertial Observers*

In studying Newton's laws of motion in introductory physics, we learn that the laws were written down for observers called inertial observers. Inertial observers are observers that are not undergoing acceleration so that they travel at constant velocity. The reason that the laws of motion were written for inertial observers is that the laws for inertial observers "look" simple. Perhaps your introductory physics course or a classical mechanics course you have taken has covered noninertial observers. Observers in noninertial reference frames either no longer find the equations of motion to be simple or they must introduce fictional forces to keep the equations of motion simple. In either case, motion is more complicated in the noninertial observers' view than in the inertial observers' view. Similarly, in special relativity, inertial observers are special observers in that the relationship between inertial observers' measurements takes a much simpler form than it does for noninertial observers.

Quantities of physical interest can be classified according to how the transformation rules must be applied to those quantities to relate measurement of those quantities by different inertial observers. The viewpoint that physicists have taken is that the physical phenomena have existence in themselves and different observers use different coordinate systems to measure the properties of those physical phenomena. This program allows the investigation of the transformation of measurements of physical quantities called four vectors and second-rank tensors. There are other types of quantities of physical interest, but this program is limited to investigating the two types above. Those types, along with scalars, which transform trivially, include most of the physical quantities you have studied in introductory courses.

The rules for transforming physical quantities by the Lorentz transformation can most easily be written down in matrix notation. Four vectors will be written, with the first component being the time component, the second component being the x spatial component, the third component being the y spatial component, and the fourth component being the z spatial component. The components will be labeled with $t, x, y,$ and z on the display rather than numerically for clarity. Within the source code of the program, they are labeled with indices 0, 1, 2, and 3, respectively. All velocities will be specified by $\beta = v/c$, where v is the velocity or velocity component of interest and c is the speed of light. In the natural units used in the program, c is identically one and this becomes trivial.

Consistently, the representation of second-rank four tensors will have the first row as the time row, the second row as the x spatial row, the third row as the y spatial row, and the last row as the z spatial row. The first column will be the time column, the second column will be the x spatial column, the third column will be the y spatial column, and the fourth column will be the z spatial column. As with vectors, in the source code of the program the rows and columns are labeled 0, 1, 2, and 3, respectively.

The Lorentz transformation matrix is then written in matrix form as

$$
\Lambda = \begin{pmatrix}
\gamma & -\gamma\beta_x & -\gamma\beta_y & -\gamma\beta_z \\
-\gamma\beta_x & 1 + \frac{(\gamma-1)\beta_x^2}{\beta^2} & \frac{(\gamma-1)\beta_x\beta_y}{\beta^2} & \frac{(\gamma-1)\beta_x\beta_z}{\beta^2} \\
-\gamma\beta_y & \frac{(\gamma-1)\beta_x\beta_y}{\beta^2} & 1 + \frac{(\gamma-1)\beta_y^2}{\beta^2} & \frac{(\gamma-1)\beta_y\beta_z}{\beta^2} \\
-\gamma\beta_z & \frac{(\gamma-1)\beta_x\beta_z}{\beta^2} & \frac{(\gamma-1)\beta_y\beta_z}{\beta^2} & 1 + \frac{(\gamma-1)\beta_z^2}{\beta^2}
\end{pmatrix},
\tag{4.1}
$$

where

$$
\beta = \sqrt{\beta_x^2 + \beta_y^2 + \beta_z^2}
\tag{4.2}
$$

and

$$
\gamma = \frac{1}{\sqrt{1 - \beta^2}},
\tag{4.3}
$$

with β_x, β_y, and β_z being the velocity components of the second reference frame with respect to the first along the x, y, and z directions, respectively.

The transformation rule for a four vector is then simply

$$
A' = \Lambda A,
\tag{4.4}
$$

where A' has components measured in the frame of an observer moving relative to the frame in which the same four vector has components measured as those of A. This transformation rule assumes that the origin and the orientation of the relative velocity vector of the two reference frames is identical in the two reference frames. To relate this to the Lorentz transformation rules learned from introductory physics texts, apply this to the location of an event in two different reference frames where the second reference frame is moving with a velocity β in the x direction relative to the first reference frame. The Lorentz transformation matrix in this case is

$$
\Lambda = \begin{pmatrix}
\gamma & -\gamma\beta_x & 0 & 0 \\
-\gamma\beta_x & \gamma & 0 & 0 \\
0 & 0 & 1 & 0 \\
0 & 0 & 0 & 1
\end{pmatrix}.
\tag{4.5}
$$

Applying the transformation rule to the event location:

$$
\begin{pmatrix} t' \\ x' \\ y' \\ t' \end{pmatrix} = \begin{pmatrix}
\gamma & -\gamma\beta_x & 0 & 0 \\
-\gamma\beta_x & \gamma & 0 & 0 \\
0 & 0 & 1 & 0 \\
0 & 0 & 0 & 1
\end{pmatrix} \begin{pmatrix} t \\ x \\ y \\ t \end{pmatrix}.
\tag{4.6}
$$

Performing the matrix multiplication on the right side of the equation results in

$$t' = \gamma t - \gamma \beta x \tag{4.7}$$

$$x' = -\gamma \beta t + \gamma x \tag{4.8}$$

$$y' = y \tag{4.9}$$

$$z' = z \tag{4.10}$$

which are the Lorentz transformation rules covered in introductory physics texts. The transformation rule for a second-rank four tensor is given by

$$M' = \Lambda M \Lambda^T, \tag{4.11}$$

where Λ^T is the transpose matrix of the Λ matrix,

$$\Lambda_{ij}^T = \Lambda_{ji}, \tag{4.12}$$

M is the matrix representing the tensor in the first reference frame and M' is the matrix representing the tensor in the reference frame related to the first by the Lorentz transformation. Maybe as somewhat of a surprise, the most familiar physical quantity that transforms as a second rank tensor is the electromagnetic field. The electromagnetic stress tensor $\mathcal{F}$ is an antisymmetric tensor. Antisymmetric means

$$\mathcal{F}_{ij} = -\mathcal{F}_{ji}, \tag{4.13}$$

which implies that the diagonal elements of this tensor are identically zero. Antisymmetry of a tensor is preserved under Lorentz transformations (see Exercise 4.1). The electromagnetic stress tensor can be written in terms of the electric and magnetic field components as

$$\mathcal{F} = \begin{pmatrix} 0 & -E_x & -E_y & -E_z \\ E_x & 0 & B_z & -B_y \\ E_y & -B_z & 0 & B_x \\ E_z & B_y & -B_x & 0 \end{pmatrix}. \tag{4.14}$$

4.3 Units

In this program, the speed of light is dimensionless and is equal to one in magnitude. To convert quantities to conventional units, select either the units for spatial components or time components of four vectors. If the unit of the spatial components is chosen, the unit of the time component is then the spatial unit divided by the speed of light in the desired unit system. For instance, if the length unit is chosen as meters, the time unit is

$$\frac{1 \text{ m}}{3 \times 10^8 \text{ m/s}} = 0.33 \times 10^{-8} \text{ s}. \tag{4.15}$$

If the time unit is chosen, the distance unit is equal to the time unit multiplied by the speed of light in the desired unit system. If the time unit is chosen as seconds, then the distance unit is

$$(1 \text{ s})(3 \times 10^8 \text{ m/s}) = 3 \times 10^8 \text{ m} = 1 \text{ light-second.} \tag{4.16}$$

As another example, note that the four vector being considered is the energy-momentum four vector. If units of MeV are chosen as the energy (time component) units, the units of momentum (spatial components) are MeV/c.

For second-rank four tensors, there are components with three different units. There are space-space components, space-time components, and time-time components. The unit of the space-time components is the speed of light multiplied by the space-space unit. The time-time component has units of the speed of light multiplied by the unit of the space-time components.

4.4   The Programs

The special relativity simulations consist of two different programs. One of these programs is named SPECREL and the second is named CLOCKS.

The program SPECREL can be considered a special relativity calculator. The program can calculate trajectories of objects and display the motion on a variety of graphs in two different reference frames simultaneously. The program can also display and calculate the components of four vectors and second-rank four tensors in two different coordinate systems simultaneously. A graphical representation of the these objects is also displayed. The first reference frame is always an inertial frame, but the second reference frame can be chosen to be a noninertial reference frame.

The program CLOCKS is program for observing the readings on clocks in relative motion and light signals passed between observers and the clocks. The program allows the specification of the trajectory of a clock through space-time and that trajectory is shown on space-time plots of two different inertial observers. The program animates the motion of the clocks and the observers in the two reference frames and shows the instantaneous readings of the clocks in each reference frame.

4.5   Details of the SPECREL Program

4.5.1   The Display

The upper half of the screen displays two graphs and buttons that control what is displayed on the graphs (Fig. 4.1). Either graph can be set to show x vs. t, x vs. y, x vs. y vs. t, or x vs. y vs. z for either of the two reference frames.

Either graph can be set to display world lines or world tubes in the background of the graphs as appropriate when t is one of the axes of the graph displayed. The current value of time in both reference frames can be run or incremented. The lower left half of the screen shows information regarding the currently displayed

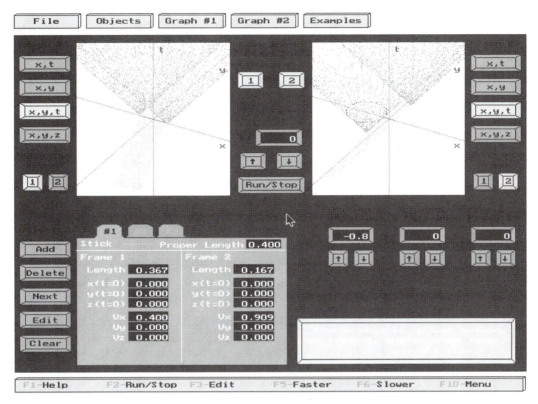

Figure 4.1: Display for the SPECREL program.

objects. When geometric-shaped objects are displayed, a deck of index cards is displayed with one card corresponding to each of the objects. The information on the cards dynamically changes with changes in the reference frame or changes in the values displayed.

The lower half of the display shows the motion of reference frame 2 with respect to reference frame 1 with some control buttons to change the values of the components of the motion. In addition, there is a small prompt display to help users to proceed.

4.5.2   Program Control

The program is controlled by selection of menu items, pressing hot keys, clicking on screen buttons, and entering requested data.

There are many buttons available on the screen to easily control the program with simple mouse button clicks. These buttons are duplicated in function by menu items for users that do not have a mouse available.

4.5.3   Menus

There are five main menu items each with several choices:

- **File**: Program information, configuration, exit program.

  — **About CUPS**: This menu item displays information about the CUPS project.

  — **About Program**: This menu item displays information about the CUPS QUANTUM program.

  — **Configuration**: This menu item allows for user selection of temporary file path directory, colors, double-click time of the mouse, and delay time. It also allows the user to view the amount of memory available on the heap.

  — **Quit**: This menu item exits the program and returns to the operating system.

- **Objects**: This menu controls the mode of the program and also contains items available from the screen buttons for users without a mouse to control the type of objects displayed.

  — **Rockets, Sticks, Flashes, etc.**: This menu item puts the program in the mode in which trajectories of objects can be investigated. Selecting this item will draw the objects in the graphs and show the information cards for the items that have been defined.

  — **Four Vectors**: This item puts the program in the mode for investigating the transformation of four vectors. Components of a four vector in two different reference frames are shown on the information card.

  — **Tensors**: This item puts the program in the mode for investigating the transformation of second-rank four tensors. Components of a second-rank four tensor in two different reference frames are shown on the information card.

  — **Add**: This item allows the user to add another object to the current list of objects.

  — **Delete**: This item deletes the object associated with the displayed index card from the object list. The objects are renumbered after the deletion to be sequentially numbered.

  — **Next**: This item advances the displayed object information to the next object in the list.

  — **Edit**: This item starts and ends the program mode to edit the entries on the displayed information card. When this item is selected the program highlights an entry on the information card shown. That entry can be changed or another entry can be selected using the **tab** key, **arrow** keys, or clicking the mouse with the cursor pointing to the desired entry. Edit mode continues until the **enter** key is pressed or the edit button is clicked on.

  — **Clear**: This menu item will remove all items from the object list and clear the graphs.

- **Graph #1**: This menu has items that control what is displayed in graph 1.

  — **xt**: This item sets graph 1 to be a graph of x versus t.

  — **xy**: This item sets graph 1 to be a graph of x versus y.

— **xyt**: This item sets graph 1 to be a graph of *x* versus *y* versus *t*.
— **xyz**: This item sets graph 1 to be a graph of *x* versus *y* versus *z*.
— **Frame 1**: This item sets graph 1 to display the objects and their trajectories in reference frame 1.
— **Frame 2**: This item sets graph 1 to display the objects and their trajectories in reference frame 2.
— **Show World Tubes**: This item sets graph 1 to display world lines or world tubes when any graph involving time as one coordinate is selected.

- **Graph 2**: This menu has items that control what is displayed in graph 2.

    — **xt**: This item sets graph 2 to be a graph of *x* versus *t*.
    — **xy**: This item sets graph 2 to be a graph of *x* versus *y*.
    — **xyt**: This item sets graph 2 to be a graph of *x* versus *y* versus *t*.
    — **xyz**: This item sets graph 2 to be a graph of *x* versus *y* versus *z*.
    — **Frame 1**: This item sets graph 2 to display the objects and their trajectories in reference frame 1.
    — **Frame 2**: This item sets graph 2 to display the objects and their trajectories in reference frame 2.
    — **Show World Tubes**: This item sets graph 2 to display world lines or world tubes when any graph involving time as one coordinate is selected.

- **Frame 2**: This menu has items that allow users without a mouse to control the motion of reference frame 2 with respect to reference frame 1 and the current time.

    — βx: This item allows setting the *x* motion of reference frame 2.
    — βy: This item allows setting the *y* motion of reference frame 2.
    — βz: This item allows setting the *z* motion of reference frame 2.
    — **Time**: This item allows the user to set a value of time.

- **Examples**: This menu has items that set up the object list and frame 2 motion to observe some of the standard examples from the study of special relativity.

    — **Pole and Barn**: This item sets the program up to display the familiar pole and barn example used to exhibit the frame dependence of simultaneity.
    — **Twin Paradox**: This item sets the program up to display the twin paradox situation.
    — **Meter Stick and Hole**: This item sets the program up to display the familiar two-dimensional motion problem of a stick passing through a hole in a plate.

4.5.4 Hot Keys

There are a number of hot keys available to help control the program:

- F1-Help: This hot key displays a help screen relevant to the current mode of the program.

- F2-Run/Stop: This hot key starts and stops incrementation of time.

- F3-Edit: This hot key puts the program into edit mode when geometric objects are displayed.

- F5-Faster: This hot key will speed up the animation if possible. It does not change the time step, but changes the delay between successive steps.

- F6-Slower: This hot key will slow down the animation. It does not change the time step, but changes the delay between successive steps.

- F10-Menu: This hot key activates the menu selection procedure.

4.5.5   Entering Values

In any mode that allows highlighting and editing a value displayed on the screen, keystrokes for input behave in a manner similar to the input on common calculators. The first legitimate keystroke deletes the old value and enters the key pressed into the display. After the initial legitimate key has been pressed, further key presses push the previous digits to the left. Exceptions to this are the minus key and the backspace key. The minus sign will toggle the leading minus sign on the entered value or the minus sign on the exponent once the e key has been pressed. The backspace key deletes characters from right to left one at a time. Any key press that is not a legitimate number entry key or control key is ignored. The enter key ends edit mode in those situations which require a confirmation that editing is complete to inform the program to begin calculations. The prompt in the lower right corner of the screen informs the user if that is necessary.

4.5.6   Structure of the Program

The program consists of the following units:

- SPECREL.PAS: This is the main program.

- SROBJ0.PAS: This module contains the structure and methods for the buttons and editable regions on the screen.

- SROBJ1.PAS: This module contains the structure and methods for the geometrical shape objects that are displayed in the **Sticks, Rockets,** etc. mode of the program.

- SROBJ2.PAS: This module contains the structure and methods for the four vectors and tensors for those respective modes of the program.

- FLASHXY.DAT: This file contains the data needed to display flashes in an x versus y graph.

- FLASHXYT.DAT: This file contains the data needed to display flashes in an x versus y versus t graph.

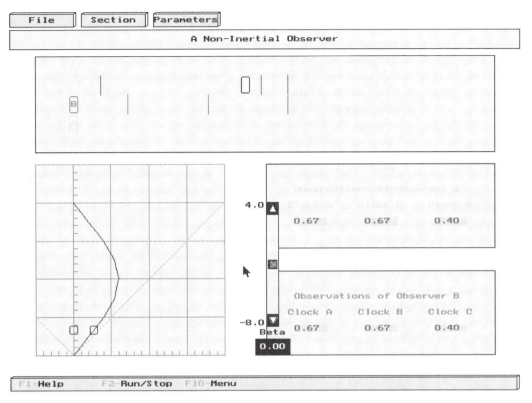

Figure 4.2: Display for the CLOCKS program.

- FLASHXYZ.DAT: This file contains the data needed to display flashes in an x versus y versus z graph.

4.6   Details of the CLOCKS Program

4.6.1   The Display

The display for the CLOCKS program is shown in Figure 4.2. The top window of the display shows rectangles labeled A, B, and C to represent clocks in reference frames A, B, and C, respectively. When the animation is run, the clocks move and emit periodic light pulses that travel across the screen at the scaled speed of light. If the sound is turned on, a beep is heard when a clock receives a pulse of light.

Below the display of the moving clocks is a space-time diagram on the left. The diagram shows the trajectories of the clocks and a coordinate grid of each of the inertial observers. At the right side of the display are two windows that show a digital display of the readings on each of the clocks for the two inertial observers. The upper window shows the observations of inertial observer A and the lower window shows the observations of inertial observer B.

4.6.2   Menus

There are five main menu items each with several choices:

- **File**:

  — **About CUPS**: This menu item displays information about the CUPS project.
  — **About Program**: This menu item displays information about the CUPS QUANTUM program.
  — **Configuration**: This menu item allows for user selection of temporary file path directory, colors, double-click time of the mouse, and delay time. It also allows the user to view the amount of memory available on the heap.
  — **Quit**: This menu item exits the program and returns to the operating system.

4.6.3 Hot Keys

There are several hot keys for controlling the program.

- F1-Help: This hot key brings up a context sensitive help screen.

- F2-Run/Stop: This hot key starts and stops the animation of the display.

- F3-Edit: This hot key puts the program into the edit mode for editing the defined objects.

- F5-Faster: This hot key increases the speed of animation incrementally with each key press if possible.

- F6-Slower: This hot key decreases the speed of animation incrementally with each key press.

- F10-Menu: This hot key activates the menu at the top of the screen.

4.7 Exercises

4.1   **Preservation of Tensor Symmetry**
From the **Objects** menu, select **2nd rank tensors**. On the display of the tensor components click on the antisymmetry box. Change components or reference frames. Does the tensor remain antisymmetric when transformed (i.e., is it antisymmetric in both reference frames)? Check the symmetric tensor box on the tensor component display. Do symmetric tensors remain symmetric under Lorentz transformations? Of course, the computer program could be cheating to keep the transformed tensors symmetric. Starting from the Lorentz transformation matrix (Eq. 4.5), the rule for second-rank tensor transformation (Eq. 4.11), and a general symmetric second-rank tensor,

$$M = \begin{pmatrix} 0 & a & b & d \\ a & 0 & c & e \\ b & c & 0 & f \\ d & e & f & 0 \end{pmatrix}, \tag{4.17}$$

show that the transformed tensor M' is also symmetric.
Repeat the demonstration for a general antisymmetric second-rank tensor

$$M = \begin{pmatrix} 0 & -a & -b & -d \\ a & 0 & -c & -e \\ b & c & 0 & -f \\ d & e & f & 0 \end{pmatrix}. \tag{4.18}$$

4.2 Invariance of Norm

The norm of a four vector is given by

$$\|v\| = v_0^2 - v_1^2 - v_2^2 - v_3^2. \tag{4.19}$$

Using the four vector objects of the program, calculate the norm of several different four vectors from the components calculated by the program for various velocities of reference frame 2. What type of quantity does the norm of a four vector appear to be? Just because it is a single number, it is easy to be tempted to answer "scalar" immediately, but scalars have an important property besides being represented by a single number. A single number also describes a component of a vector or a component of a second-rank tensor. Does the norm of the four vector have the appropriate behavior of a scalar when transformed to different inertial reference frames?

4.3 Proper Time

The norm of the position four vector is called the proper time, τ. Verify that $dt = d\tau$ only along a path through space-time that is at constant position in a given inertial reference frame. This implies that a clock measure $d\tau$ for the path through space-time that the clock follows. The norm of a vector is a scalar, so the reading on the clock must be the same according to all observers.

4.4 Simultaneity

To demonstrate the frame dependence of simultaneity, add the following objects in frame 1: a stick of length 0.2 with $x(0)$, $y(0)$, $z(0)$, v_x, v_y, and v_z all equal to zero; a flash with $x(0) = -0.1$, $y(0)$, $z(0)$, and $t = 0$; and a flash with $x(0) = +0.1$, $y(0)$, $z(0)$, and $t = 0$. When viewed in any graph in frame 1, the stick is stationary and the two flashes occur simultaneously from the ends of the stick. Select various directions and velocities of motion for reference frame 2. Observe how the relative time of occurrence of the two flashes depends on the direction and magnitude of the velocity of the reference frame.

4.5 Velocity Transformation

Introductory texts often derive the transformation of velocity between different inertial reference frames by transforming dx and dt and then taking the quotient. Similarly the y and z velocity components are calculated by transforming dy and dz and dividing by the transformed dt. As velocity is a vector quantity in nonrelativistic physics, it might be expected that there is a corresponding four vector to velocity. Use the program to verify that

$$v_4 = (0, v_x, v_y, v_z) \tag{4.20}$$

does not transform as a four vector. Construct a vector with a zero time component, and non-zero x, y, and z components and view the components in different reference frames to see that the time component does not remain zero and the other components do not transform as expected from the velocity transformation rules derived in introductory physics. Is there a four vector quantity that has spatial components closely related to the nonrelativistic velocity? It turns out that the four vector defined by

$$v_4 = (\gamma, \gamma v_x, \gamma v_y, \gamma v_z) \tag{4.21}$$

is a four vector. Try to verify this by setting up a four vector in reference frame 1 with the time component equal to 1 and the spatial components set to some non-zero values. Verify that the values of the components are correctly given for frame 2 for various velocities of frame 2. If you think carefully about the energy-momentum four vector, you could easily verify that the definition above does indeed define a four vector. Verify that the velocity expression above is equivalent to $dX/d\tau$, where X is the four vector that represents the coordinates of the object.

4.6   Acceleration Transformation

The acceleration of an object measured by an observer is the change in velocity divided by the change in time, dv/dt. In Exercise 4.5 it was shown that the four velocity, which is the derivative of the four position with respect to the scalar $d\tau$, is itself a four vector. We should expect that the four acceleration is the derivative of the four velocity with respect to $d\tau$ and that this is also a four vector. Show that the four acceleration is given by

$$a_4 = \left(\gamma \frac{d\gamma}{dt}, \gamma \left(\frac{d\gamma}{dt}\frac{dx}{dt} + \gamma \frac{d^2x}{dt^2} \right), \right.$$
$$\left. \gamma \left(\frac{d\gamma}{dt}\frac{dy}{dt} + \gamma \frac{d^2y}{dt^2} \right), \gamma \left(\frac{d\gamma}{dt}\frac{dz}{dt} + \gamma \frac{d^2z}{dt^2} \right) \right). \tag{4.22}$$

Note that the time component of the four acceleration in the rest frame is zero, but not in any other frame. Define a four vector in frame 1 with a time component equal to zero and the x component equal to one. Verify that the four vector transforms by the program according to the rule established above.

4.7   Orthogonality of Four Velocity and Four Acceleration

Verify that the four velocity and four acceleration are orthogonal.

References

1. Resnick. R., Halliday, D. *Basic Concepts in Relativity and Early Quantum Theory*. New York: Macmillan Publishing Company, 1992.

2. Taylor, E., Wheeler, J. *Spacetime Physics*. San Francisco: W. H. Freeman and Company, 1963.

3. Rindler, W. *Essential Relativity.* New York: Springer-Verlag, 1977.

4. Wald, R. *General Relativity.* Chicago: The University of Chicago Press, 1984.

5. Jackson, J. *Classical Electrodynamics,* 2nd ed. New York: John Wiley & Sons, 1974.

5

Laser Cavities and Dynamics

Michael J. Moloney

5.1 General Background

Lasers involve light interacting with atoms in a cavity to produce a coherent standing-wave pattern. Two light fields are important. One light field "pumps" the atoms to higher states. A second light field forms a standing wave pattern, stimulating excited atoms to emit in phase, and thus to build the standing wave to levels which support significant coherent output through one end-mirror of the laser cavity.

Each atom has a set of energy levels; each energy level has properties which depend on the details of that atom. In this chapter, energy levels will be identified numerically (level 1, level 2, etc.) in order of increasing energy ($E_2 > E_1$, etc.).

Laser output radiation requires an excited state (say, level 2) of the atom which has a relatively long lifetime. This state can de-excite to a lower state (level 1) by emitting a photon. For lasing, the level 2 population must be made to exceed the level 1 population. This is accomplished by subjecting the atoms to a radiation field which excites ground-state atoms to an excited state (level 3). This "pumping" process creates a large population in level 3, which rapidly decays to level 2. Atoms accumulate in level 2, where they are stimulated to decay to level 1, giving up photons which add in phase with the cavity electric field.

This simulation examines a three-level laser which utilizes the ground state (level 0) of an atom, along with levels 1 and 2. Pumping occurs from level 0 to level 2; this is roughly equivalent to a four-level scheme where pumping is done from level 0 to level 3 which rapidly decays to level 2.

5.2 "Blackbody" or "Temperature" Radiation

A body emits radiation characteristic of its temperature; as a body becomes hotter, the radiation it emits shifts to shorter wavelengths. We are familiar with seeing the colors change as a body is heated; from dull red to dull orange to white, and possibly to bluish white as the body's temperature increases. This is indeed "temperature" radiation, although it is usually called "blackbody" radiation.

In a cavity the radiation density ρ is the energy per unit volume and per unit frequency. This density is independent of the nature of the walls of the cavity. A simple thought experiment due to G. R. Kirchhoff (1859) shows this to be true. Two cavities (of different wall materials) are at the same temperature T, and are connected by a tube which only permits light of frequency f to pass through it. If the radiation density ρ depended on the wall material, one cavity would send more energy steadily to the other, and spontaneously destroy the temperature equilibrium between them.

The curve of cavity radiation was measured at the turn of the century, and a formula for it was found by Planck[1] :

$$\rho = \frac{dU}{df} = \frac{8\pi h f^3}{e^{E/kT} - 1},$$ (5.1)

where h is Planck's constant, k is Boltzmann's constant $(1.3805 \times 10^{-23} \text{J}/K)$, U is the energy per volume, and f is the frequency of the light. $E = hf$ is the energy of each photon, or "chunk" of light, an idea which was established after Planck did his original derivation with oscillators in the walls of the cavity.

In what follows, hf will represent the energy between two distinct energy levels in an atom. We will write it as

$$E_{mn} = E_m - E_n = hf_{mn},$$ (5.2)

and the radiation density ρ_{mn} will be written

$$\rho_{mn} = \frac{8\pi h f_{mn}^3}{e^{hf_{mn}/kT} - 1}.$$ (5.3)

5.3 Spontaneous and Stimulated Emission

For two energy levels, E_1 and E_2, of an atom in a cavity at temperature T, it was established by Boltzmann that the populations of atoms (n_1, n_2) in the energy levels are related by

$$\frac{n_2}{n_1} = e^{-(E_2 - E_1)/kT}.$$ (5.4)

Atoms can be excited into various levels and also decay into lower levels. For atoms not in contact, radiation must "stimulate" atoms in lower levels to higher levels, while atoms in higher levels could either spontaneously decay into lower levels or be stimulated to decay by cavity radiation. This process must work out so that equilibrium is maintained between populations of all energy levels. The rate of level 2 atoms going to level 1 per unit time is

$$R_{2\to 1} = A_{21}n_2 + B_{21}\rho_{21}n_2,$$ (5.5)

where the first term represents the atoms spontaneously decaying from level 2 to level 1, and the second represents atoms stimulated to decay by the radiation density ρ_{21} at frequency $f_{21} = (E_2 - E_1)/h$. The coefficients A and B were introduced by Einstein,[2] and are often called the spontaneous and stimulated emission coefficients. The rate of level 1 atoms going to level 2 is

$$R_{1 \to 2} = B_{12}\rho_{21}n_1, \qquad (5.6)$$

where ρ_{21} is the radiation density at frequency $f_{21} = (E_2 - E_1)/h$ (see Eq. 5.3), while A_{21}, B_{21}, and B_{12} are constants depending on the type of atom we have. The *net* flow rate of atoms from level 2 to level 1 is

$$F_{21} = R_{2 \to 1} - R_{1 \to 2} = \text{net atoms/s from level 2 to level 1.} \qquad (5.7)$$

Einstein's classic 1917 paper,[2] in which these equations were first set forth, also established that $B_{mn} = B_{nm}$. If an atom contains only two levels, the rate of population change of level 2 is

$$-\frac{dn_2}{dt} = F_{21} = A_{21}n_2 + B_{21}\rho_{21}(n_2 - n_1). \qquad (5.8)$$

At thermal equilibrium, n_2 is constant (and equilibrium must be maintained between each pair of levels; all $F_{mn} = 0$), so we can set $dn_2/dt = 0$ and solve Eq. 5.8 for n_2/n_1:

$$\frac{n_2}{n_1} = \frac{\rho_{21}}{A_{21}/B_{21} + \rho_{21}}. \qquad (5.9)$$

Using Eqs. 5.4 and 5.9, n_2/n_1 can be eliminated:

$$X_{21} = B_{21}\rho_{21} = \frac{A_{21}}{e^{(E_2 - E_1)/kT} - 1}. \qquad (5.10)$$

Equation 5.10 will be used to eliminate $B_{21}\rho_{21}$ from the upcoming rate equations. A further substitution, from Eq. 5.3, shows that A_{21} and B_{21} are related through the energy difference between level 1 and level 2:

$$B_{21} = \frac{A_{21}}{8\pi h f_{21}^3}. \qquad (5.11)$$

Two important parameters of the atoms in the laser may be adjusted in this simulation; the A_{mn} ("decay constants," to be discussed in the next paragraph), and the atomic energy level differences E_{mn}. Einstein B_{mn} coefficients follow from the A_{mn} and E_{mn} according to Eq. 5.11, so both Einstein A and B coefficients are tied up with the details of atomic structure, but not with the temperature.

For most atoms at room temperature, $E_2 > E_1$ and $e^{(E_2 - E_1)/kT} \gg 1$, meaning that $A_{21} \gg B_{21}\rho_{21}$ at thermal equilibrium. From this we can show that the Einstein A coefficient is in effect a "decay constant," or "reciprocal lifetime." If n_2 were to increase greatly, due to a burst of pumping radiation which is then shut off, the right-hand side of Eq. 5.8 would be dominated by $A_{21}n_2$, and Eq. 5.8 would become

$$\frac{dn_2}{dt} = -A_{21}n_2. \qquad (5.12)$$

While $n_2 \gg n_1$, the decay of n_2 would be described by $n_2(t) = n_2(0)e^{-A_{21}t}$. Large A_{21} values mean a short lifetime and rapid decay, while small A_{21} values mean a long lifetime and slow decay.

5.4 Laser Thermal Equilibrium

Assume atoms with only three energy levels E_0, E_1, and E_2, having populations n_0, n_1, and n_2 in each level. The total population of atoms is N. The steady-state equations for the populations of levels 0, 1, and 2 at thermal equilibrium (after using Eq. 5.10 and its analogs to replace B_{mn} terms) are

$$n_0 + n_1 + n_2 = N = n_{\text{total}}, \tag{5.13}$$

$$dn_0/dt = +F_{20} + F_{10} = +[R_{2\rightarrow0} - R_{0\rightarrow2}] + [R_{1\rightarrow0} - R_{0\rightarrow1}] = 0, \tag{5.14}$$

$$dn_1/dt = +F_{21} - F_{10} = +[R_{2\rightarrow1} - R_{1\rightarrow2}] - [R_{1\rightarrow0} - R_{0\rightarrow1}] = 0, \tag{5.15}$$

and

$$dn_2/dt = -F_{21} - F_{20} = -[R_{2\rightarrow1} - R_{1\rightarrow2}] - [R_{2\rightarrow0} - R_{0\rightarrow2}] = 0, \tag{5.16}$$

where

$$F_{20} = A_{20}n_2 + X_{20}(n_2 - n_0), \tag{5.17}$$

$$F_{10} = A_{10}n_1 + X_{10}(n_1 - n_0), \tag{5.18}$$

and

$$F_{21} = A_{21}n_2 + X_{21}(n_2 - n_1). \tag{5.19}$$

In steady state, the populations n_0, n_1, and n_2 have constant values, given the energy levels, Amn, and the temperature, so that the rate of increase of any level with time must be zero, as shown by Eqs. 5.14–5.16. Well-known populations emerge from the solution to these equations:

$$n_1 = n_0 e^{-E_{10}/kT}, \tag{5.20}$$

$$n_2 = n_0 e^{-E_{20}/kT}, \tag{5.21}$$

$$N = n_{\text{total}} = n_0 + n_1 + n_2, \tag{5.22}$$

and

$$n_0 = \frac{N}{1 + e^{-E_{10}/kT} + e^{-E_{20}/kT}} = \frac{N}{Z}. \tag{5.23}$$

Thermal equilibrium populations may be summarized by[5]

$$n_m = \frac{Ne^{-E_{m0}/kT}}{Z}, \tag{5.24}$$

where $m = 0, 1$, or 2, and

$$Z = 1 + e^{-E_{10}/kT} + e^{-E_{20}/kT}. \tag{5.25}$$

Populations are always smaller in higher energy levels than in lower ones at thermal equilibrium.

5.5   Conditions for Lasing

Lasing between level 2 and level 1 means that atoms have built up in level 2 and are stimulated to drop to level 1 by an existing radiation field in the laser cavity. Photons emitted from this decay add to the coherent radiation field.

The second term of Eq. 5.8 deals with the net flow of atoms from level 2 to level 1 by stimulated emission, and shows that a net flow of atoms from level 2 to level 1 requires more atoms in level 2 than in level 1. Two things are needed to achieve this "population inversion."

- 1. Atoms must be rapidly resupplied to level 2.

- 2. The lifetime of atoms in 2 for spontaneous decay to level 1 must be fairly long so they accumulate and "stay" in level 2.

The second requirement is achieved only if A_{21} is small. This is an atomic property; an energy level must be chosen which has a long lifetime. In the model, one adjusts A_{21} downward.

The requirement of resupplying atoms to level 2 is met by "pumping" from level 0 to level 2, often via an additional radiation beam. When pump power is turned on, it enhances the thermal equilibrium radiation density ρ_{20}, which stimulates atoms to make transitions between n_0 and n_2. In Eq. 5.17, X_{20} becomes PX_{20}, because pumping enhances the thermal equilibrium radiation density ρ_{20} by a factor of P. Solving for the populations with pumping is thus no more difficult than at thermal equilibrium.

(In practice, the "three-level" lasing scheme outlined above really involves a fourth level. Pumping actually takes place from level 0 to level 3, above level 2, which has a very short lifetime for decay into level 2.)

With the same pumping going on from level 0 to level 2, we could also have lasing if there were a long lifetime between level 1 and level 0. That way, atoms would be pumped to level 2, decay quickly to level 1, then accumulate so as to create the possibility of lasing as they are stimulated to drop to level 1. This scheme is designated as "1 $\rightarrow$ 0" lasing.

One of the suggested exercises compares the coherent light output of 2$\rightarrow$1 lasing to that of 1$\rightarrow$0 lasing with the same pumping radiation field between levels 0 and 2. These two lasing schemes will be seen to have dramatically different optical outputs. (Each type of lasing scheme would occur in a different laser.)

5.6   The Lasing Threshold and Coherent Radiation Density

When "2$\rightarrow$1" lasing begins, one or more decays from level 2 to level 1 set up a coherent radiation density W which will grow with every pass back and forth between the laser mirrors . But as W grows, it starts depopulating level 2 through stimulated emission. W grows and n_2 decreases until steady levels of W and n_2 are present. The gain-per-pass due to transitions from level 2 to level 1 is then matched by loss-per-pass due to mirror transmission, diffraction losses, scattering from atoms, etc. The rate of change of W with time is given by

$$dW/dt = c_{gain}W(n_2 - n_1) - c_{loss}W, \text{ or} \tag{5.26}$$

$$dW/dt = KW, \tag{5.27}$$

where

$$K = c_{gain}(n_2 - n_1) - c_{loss}. \tag{5.28}$$

Thus, the coherent density W grows or decays exponentially ($W = W_0 e^{Kt}$), depending on the value of K. The c_{gain} term represents growth of the coherent density due to stimulated transitions (c_{gain} contains B_{21}) from level 2 to level 1. The c_{loss} term represents losses in coherent intensity due to miscellaneous effects and also transmission through one of the end mirrors of the cavity. The lasing threshold with no mirror transmission is taken to be such that $n_2 - n_1$ equals 5% of the total number of atoms present, N:

$$n_2 - n_1 = 0.05N = \left[\frac{c_{loss}}{c_{gain}} \right] \text{ no mirror transmission}. \tag{5.29}$$

Miscellaneous cavity losses are taken to be 2% of the coherent intensity W when there is no mirror transmission (i.e., $c_{loss} = 0.02$). When K is negative, losses exceed gains, and W decays to zero. When K is positive, W begins to grow exponentially, but as it does $(n_2 - n_1)$ decreases until K exactly equals zero, and the value of W becomes constant. As the pump power increases, the population level difference will increase, and so will W.

When $n_2 - n_1$ exceeds the nominal threshold of $0.05N$, 2→1 lasing begins. A coherent radiation density then appears due to lasing, which stimulates transitions between level 2 and level 1. Expressing this coherent density as a multiple W of the thermal equilibrium radiation density ρ_{21} means that the X_{21} term becomes WX_{21} in Eqs. 5.14–5.19. As just discussed, the difference between n_2 and n_1 is fixed at $0.05N$ (with no mirror transmission) due to the requirement that $dW/dt = 0$. This constraint is applied to determine W, once it has been determined that $n_2 - n_1$ will exceed $0.05 N$. When there is transmission through the end mirror, a larger population difference is required for lasing. At a large enough mirror transmission, the pump rate may be insufficient to sustain lasing.

All of the discussion of 2→1 lasing can be applied to 1→0 lasing, if X_{10} replaces X_{21} and $n_1 - n_0$ replaces $n_2 - n_1$.

5.7 Gaussian Beams

The laser cavity is formed by two spherical mirrors and contains an electromagnetic standing wave in the form of a "Gaussian" beam which must satisfy the boundary conditions at each mirror.[4] A Gaussian beam is a circularly symmetric wave whose energy is confined about its axis (the z-axis, through the centers of the mirrors) and whose wavefront normals are paraxial rays.

In this simulation, the user may send rays back and forth in the cavity, visually outlining the shape of the Gaussian beam. Tools in the simulation allow a quantitative comparison between the "envelope" of the ray diagram and the properties of Gaussian beams given below. This permits the user to demonstrate a connection between pure ray tracing and an electromagnetic resonant wave.

For an axially symmetric light beam, with z as the direction of propagation, and r as the coordinate transverse to the z-direction, the light intensity (apart from phase factors) may be written[4]

$$I(r,z) = I_o(w_o/w(z))^2 e^{-2(r/w)^2}, \tag{5.30}$$

where the width w in the transverse direction is

$$w = w(z) = w_o(1 + \frac{z^2}{z_o^2})^{\frac{1}{2}}. \tag{5.31}$$

When r $=$ w, the beam intensity is $1/e^2$ of its value along the z-axis. The minimum value of $w(z)$ is w_o, which occurs when z $=$ 0. The parameter z_o is known as the "Rayleigh range," and may be seen from Eq. 5.31 to be the z-value where the beam width w is $\sqrt{2}w_o$. Since all z-values are referenced to z $=$ 0 at the beam waist, z_o is the distance from beam waist to a point where the beam width is a factor of $\sqrt{2}$ greater than at the waist.

For values of r where $r^2 << z^2$, the wavefronts of a Gaussian beam are essentially spherical, with z $=$ 0 as the center of the sphere. The radius of curvature R of the spherical wavefront is given by

$$R = \frac{1}{z}(z_o^2 + z^2) = z(1 + \frac{z_o^2}{z^2}). \tag{5.32}$$

Because z_o is real for a Gaussian beam, and R has the same sign as z, when both mirrors are convex to the right (the center of curvature is to the left of the mirror vertex), z and R will be positive at both mirrors, and the beam "waist" will lie to the left of both mirrors.

When both mirrors bulge outward away from the interior of the cavity (as shown in Fig. 5.1), z and R will be negative at the left-hand mirror, while z and R will be positive at the right-hand mirror. (z will be zero at the beam waist between the mirrors.) For this geometry, the user may insert a ray in the cavity and observe the beam envelope (if the cavity is stable). Using the **BmWidth** feature, one may move the vertical bar to find the "raw" z-values (one might think of them as z'-values) of each mirror and the beam waist. The z' value of the beam waist is subtracted to give mirror z-values. Equation 5.32 is used to find z_o independently at each mirror; these two values should agree rather closely. Having determined z_o, the user may verify the variation of beam waist according to Eq. 5.31.

5.8   Details of the Program

5.8.1   Laser Operation Screens

One may study the details of laser operation by manipulating energy levels, temperature, pump power, and other parameters via sliders on each of two types of screen. For each type, one may study "2→1" lasing, or "1→0" lasing.

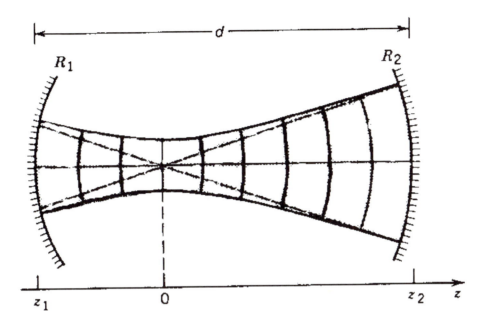

Figure 5.1: A Gaussian beam in a cavity of mirrors R_1 and R_2.

The first screen type shows the populations of each energy level, coherent cavity intensity, and output laser intensity in bar graph form. One has a visual sense of the population changes via the changing bars. Lasing shows up vividly with a bright bar, and when mirror transmission is not zero, coherent output may also be seen in bar form. Qualitative changes in populations and coherent intensity due to changes in laser operating parameters are in the forefront of this display. Numerical values of populations are also shown.

The second screen type shows the flow rate (atoms/s) between each pair of energy levels. Both spontaneous and stimulated emission are shown, in the form of bars with arrows (the width of the bar is proportional to the flow rate). All the same laser operating parameters may be adjusted as in the first screen type, but now the flow rates between levels are visually at the forefront, rather than populations and coherent intensities.

5.8.2   Laser Cavity Stability

This area of the program lets the user change mirror geometries in the cavity to verify stable and unstable lateral confinement of the beam in the laser cavity. It is interesting that the launching of paraxial rays produces a picture entirely consistent with that of electromagnetic theory for Gaussian beams.

The user may insert rays which are launched randomly along the axis, or may launch rays at specific angles to the horizontal. Rays are tracked through a number of bounces to a user-specified maximum number of bounces. If a ray reaches an angle of around 30° to the axis, it is presumed to be going out of bounds, and no more bounces are followed.

One purpose for this part of the simulation is to let the user explore the mirror radius and separation values which will create stable cavities. A second purpose is to permit (where both mirrors bulge outward from the cavity) a study of Gaussian beam properties. Gaussian beams have a wide range of applicability in many optical systems other than laser cavities, so it is worthwhile to become familiar with some of their properties, and to recognize that the "envelope" of cavity paraxial rays generates a thoroughly accurate Gaussian beam shape.

5.8.3   Beam Width (for Gaussian Beam Analysis)

After a ray has been inserted and bounced around in a stable cavity, the pattern of rays outlines an envelope of the Gaussian beam. A vertical bar is available, whose axial location z' and length of $2r$ (r was defined prior to Eq. 5.30) may be adjusted via mouse-dragging or keystrokes. z'-values must be converted to z-values (for use in Eqs. 5.31 and 5.32) by subtracting the z'-value of the beam waist. With z and r values, the user may calculate z_o. With a value of z_o, the beam width $w(z)$ can be calculated and compared to the measured width at different z-values.

To more closely fulfill the condition that $r^2 << z^2$, a magnification in the y-direction (r-direction) is offered. This magnifies the y-coordinate by 10, with z left alone. One may then launch beams at much smaller angles, and get a clear picture of the beam shape on the screen. This will provide a more precise evaluation of the Gaussian beam properties. The user must keep in mind that the screen value of beam transverse dimensions must be divided by 10 when the $\times 10$ magnification is selected. When $\times 10$ magnification is selected, *the mirrors are not replotted*; they would look quite flat if replotted, since only their central portions would be seen.

5.8.4   Two-Level and Three-Level Lasing

This area explores the detailed relationships between atomic energy level spacings, transition probabilities between levels, temperature, and pump rate in three separate atomic systems. The simplest system has only two levels, and pumping may occur between the ground state and single excited state. This system cannot lase, because even with massive pumping (see the first exercise) the population of the excited state cannot exceed that of the ground state.

The other two systems contain three levels, with pumping possible between the ground state (level 0) and the upper level (level 2). In one system, lasing is possible between level 2 and level 1, and in the other system, lasing is possible between level 1 and level 0. The user may vary any of the Einstein A coefficients, the spacing of the energy levels, the pump rate, and the temperature. The transmission coefficient of one mirror is variable, so that the user may observe the effect on the coherent intensity in the cavity, as energy is removed to form an output beam.

An arbitrary nominal value of $N = 400$ total atoms has been assigned. This was done partly to avoid use of scientific notation on screen displays.

5.9   *Suggestions for Further Study*

This simulation benefitted from a number of sources. Reference 4 is an excellent place to begin a study of Gaussian beams. The author found reference 6 particularly helpful and recommends it for further study. References 7 and 8 point toward experimental work in the context of the rate equations. They also contain valuable references to additional work.

5.10 Exercises

5.1 **Lack of Lasing With Two Levels**

    a. Use the program to show that a two-level system cannot lase. You need to show that the population of level 1 never exceeds that of level 0 for any choice of parameters for the two-level scheme. Record and turn in a summary of all parameters (including temperature, pump rate, A_{mn}) in the two-level scheme for four different sets of conditions.

    b. Prove using equations that lasing is impossible for a two-level system. One way to do this is to write down an equation like Eq. 5.8 for a two-level system with a pump rate P. A similar idea is to remove the unwanted terms from Eq. 5.14 or Eq. 5.16 when only two levels are present. Show that even with a large pump rate P, the ratio n_1/n_0 can never exceed 1.

5.2 **Verifying the Boltzmann Distribution**

    a. Check that the relative level populations $n_0:n_1:n_2$ satisfy the Boltzmann distribution. Under **Lasers** on the top menu, bring up the simulation with lasing between levels 2 and 1, and make sure the temperature is set close to 300 K. Adjust the energy levels (and possibly the temperature) so that the ratio of $n_0:n_1:n_2$ is 3:2:1. Record and turn in all the energy level differences, and the temperature (which should be 300 K).

    b. Set the temperature for 400 K repeat part a, leaving the energy levels the same. Record your results for the ratio $n_0:n_1:n_2$

    c. Make use of the fact that $kT = 0.0259$ eV when $T = 300$ K, and calculate the n_2/n_1 ratio from the formula $\exp((E_1 - E_2)/kT)$. Does this compare to the ratio you found in part a? Repeat for the ratio n_1/n_0 at 300 K and compare it to what you found on the screen.

    d. Discuss the results of part b without doing any calculations. Describe how the increased temperature changes the ratio of the energy levels. [Hint: Eq. 5.4 should be your guide.]

5.3 **The Effect of A_{mn} on Equilibrium Populations**

    a. Based on your knowledge of atomic energy levels and populations, state what effect the Einstein A_{mn} spontaneous emission coefficients have on atomic level populations at thermal equilibrium. Will the population of a level become very large if A_{mn} is very small? Explain your reasoning.

    b. How can you use the simulation to test the effect of A_{mn} on populations? Carry out the test, and explain carefully what you did. If you tried some things which you thought would work, and they didn't, mention these as well. Are results the same for two-level and three-level simulations?

5.4 A_{mn} as a Reciprocal Lifetime

a. Bring up the simulation with lasing between levels 2 and 1. Increase the pump rate (note that it's the log of the pump rate, so the pump rate is rising a lot faster than it looks) until the populations of levels 2 and 1 visibly increase. Write down your prediction for what should happen when the A_{21} coefficient is decreased. Keep in mind that this is responsible for spontaneous decays from level 2 to level 1.

b. Now decrease the A_{21} coefficient. (The log of A_{21} is represented, and decreasing A_{21} means that the log should become more negative than it was.) Describe what happens to the populations of levels 2 and 1. Explain why you think this is going on. (If this agrees with your reasoning in part a, you won't have much to add to what was said in part a.)

c. Now slowly decrease the temperature, and describe what happens to the populations of levels 1 and 2. Show that this behavior is predicted by the Boltzmann distribution, by writing down and interpreting the equations.

5.5   Effect of Pump Rate on Population in Two-Level Simulation

a. Verify that the populations are affected by pump rate in a way you understand. Bring up the two-level simulation under **Laser**. Leave temperature and energy level differences at their default values. Adjust the pump rate until the population of the upper level is half that of the lower level, as well as you can judge it by eye. Write down the log of the pump rate. Deduce the pump rate from this, keeping in mind that the log is to the base 10. Record the temperature, and the energy difference between lower and upper levels, in eV.

b. Modify Eqs. 5.14–5.16 for pumping in a two-level system, by eliminating level 1 entirely. (Hint: Eq. 5.15 will disappear entirely, n_1 will equal zero, and so will all A_{mn} or B_{mn} when $m = 1$ or $n = 1$. The upper level will be level 2 and the lower level will be level 0.)

c. Solve these equations for n_2/n_0. Eliminate A_{20}, B_{20}, and ρ_{20} using an equation equivalent to Eq. 5.8.

d. Solve for P by setting $n_2/n_0 = 0.5$. Compare your answer to that in part a. (Don't forget that kT at 300 K is 0.0259 eV.)

5.6   Effect of Temperature and Energy Level on Population

a. Bring up the two-level simulation and leave the pump rate at its default value [log(pump rate) = 0, pump rate = 1.0]. Adjust both energy level separation and temperature until the population of the upper level is half that of the lower level, as well as you can judge it by eye. Record the value of temperature and energy level separation when this occurred.

b. Solve the two-level equations you found in the previous exercise for n_2/n_0. Use $P = 1.0$ and the values for energy level separation and temperature for part a. How close is your answer to 0.5?

5.7 "Low" Pump Rate and Three-level Populations

This is an exercise in estimating important quantities in a complex set of equations and making approximate predictions without solving the equations completely. Consider a three-level atom with equally spaced energy levels and equal Einstein coefficients A_{20}, A_{21}, and A_{10} (corresponding to the default values in the simulation).

a. For equal flows in and out of level 1, use Eq. 5.15 to show that

$$n_2 - n_1 \approx n_0 e^{-E_{10}/kT} \tag{5.33}$$

until n_1 starts being appreciable with respect to n_0. Include all steps in your reasoning; be specific about why a given term can be neglected. (This result says that at low pump rates, the difference between n_2 and n_1 will be nearly constant.)

b. Show that

$$n_2 \approx \frac{P}{2} n_0 e^{-E_{20}/kT}. \tag{5.34}$$

(Hint: Eqs. 5.13 and 5.16 will be useful.) Clearly show your work in arriving at the answer.

5.8 Testing "Low" Pump Rate Predictions

Begin with "default" values for $2 \rightarrow 1$ lasing. Leave A_{20}, A_{21}, and A_{10} at their initial settings.

a. Test the predictions of the previous exercise, part a. Choose three widely different P-values; record your results for n_2, n_1, and P in a short table. Comment on the validity of the approximation.

b. Test the predictions of the previous exercise, part b. Choose four widely different P-values; recording n_2 and P-values. Within what percentage do your results agree with Eq. 5.34 if P is less than 10,000?

5.9 Population Decrease in $1 \rightarrow 0$ Lasing

In the text discussion of $2 \rightarrow 1$ lasing, it is pointed out that the populations n_1 and n_2 both increase with pump power beyond the lasing threshold.

a. Carry out the $1 \rightarrow 0$ lasing simulation and discover how levels n_1 and n_0 behave with increasing pump power beyond lasing threshold. Describe what behavior you found. Include a sketch of population vs. pump power beyond lasing threshold.

b. Explain in qualitative terms why this behavior is different than is seen in $2 \rightarrow 1$ lasing.

5.10 Thermal Equilibrium and the Lasing Threshold

Bring up the $2 \rightarrow 1$ lasing simulation. Set A_{21} to be smaller than A_{20} and A_{10} (this makes the lifetime of the $2 \rightarrow 1$ transition rather long, and means that atoms will accumulate in level 2). Raise the temperature to 800 K. Lower E_{21} and E_{10} to about 0.15 eV each. Go to **Do Plots** and select the plot of populations vs. pump rate. This plot goes from the present pump rate (1.0 times the thermal equilibrium density) to ten times this pump rate. The numbers on the x-axis are scaled, and at this low pump rate are not plotted. You should see the population of level 2 rising (due to its long lifetime) and that of level 1 falling. Estimate the populations at the left-hand edge of the plot (here the pump rate is just what occurs due to blackbody radiation at thermal equilibrium). Write down the starting values of level 1 and 2 populations, as well as you can estimate them. Go back and increase the pump rate by a factor of three [increase the log(pump rate) by about 0.5].

5.11   Lasing Threshold, Including Temperature Effects

a. Set A_{21} smaller than others, and set T = 350 K. Increase the pump rate until a coherent intensity "bar" shows up on the screen. Increase the pump rate. Describe what happens to n_1 and n_2 as the pump rate is increased.

b. Vary the temperature and determine the impact of temperature on lasing threshold. (Can you make coherent intensity start or stop at a given pump rate by adjusting the temperature?). Write down an explanation of how temperature changes should affect the lasing threshold.

5.12   Cavity "leakage" to Produce an Output Beam

a. Predict what will happen to the coherent density W and to the output beam as the mirror transmission T is increased from zero. (The output beam is proportional to both T and W, and is taken to be just T^*W.)

b. Go to the 2→1 simulation and set conditions for coherent beam in the cavity. Then gradually increase the mirror transmission and observe the effects on n_2, n_1, W, and the output beam. (On the top menu, under **Do Plot**, you can select a plot of output intensity versus mirror transmission, to see some of the same things.) Carefully describe what happens to n_1 and n_2 as the mirror transmission T is increased. Explain this behavior in terms of what was said about c_2/c_1 earlier.

5.13   Population Changes With Mirror Transmission

a. Arrange for lasing between level 2 and level 1. Use the slider to slowly increase the mirror transmission. Notice what happens to the populations of level 1 and level 2, and to the coherent intensity W. Without changing any settings, leave this area and do a plot of population versus mirror transmission. Write down the parameters with no mirror transmission, and say what happens to the populations n_2 and n_1 as mirror transmission increases.

b. Explain this behavior; what is causing n_2 and n_1 to respond in this fashion as mirror transmission increases? (Hint: Study the text prior to Eq. 5.29.)

5.14   Output Changes With Mirror Transmission

Arrange for lasing between level 2 and level 1. Use the slider to slowly increase the mirror transmission. Notice what happens to the populations of level 1 and level 2, and to the coherent intensity, W. Without changing any settings, leave this area and do a plot of coherent intensity versus mirror transmission. You should find from both of the above that the laser output goes through a maximum and then declines. Explain this behavior; what opposing factors cause the laser output to go through a maximum, then decline? (Hint: Study the text prior to Eq. 5.29.)

5.15   **Lasing Population Difference With Mirror Transmission**

a. The nominal population difference for the onset of lasing is taken as $0.05N$ when there is no mirror transmission (N is the number of atoms in the simulation). Write an expression for the population difference required for lasing onset for any value of mirror transmission, t_{mirror}. It should reduce to the correct value with no transmission, and be consistent with other assumptions made in the text.

b. Carry out a simulation which will allow you to check your formula from part a. Be specific on the details of the simulation—type of screen, parameters used, results observed.

5.16   **Mirror Transmission Versus Excess Population**

Find the mirror transmission which would require an excess population for lasing equal to 20% of the total atoms present ($0.20N$).

5.17   **Effect of A_{21} on Coherent Output**

a. Use a **Flow Rate** screen for 2→1 lasing. Set log A_{21} to about -7.5 and mirror transmission to about 0.01. Go above the lasing threshold. Now move A_{21} lower and observe the populations and flow rates. Record your observations.

b. Without changing any parameters, go to the other 2→1 lasing screen, where populations and coherent intensities are shown. Vary log A_{21} from -7.0 to -10.0 and observe the changes in populations and coherent intensities. Record your observations.

c. Explain how varying A_{21} could produce dramatic changes on one screen without producing any changes at all on the other. This is a qualitative question; it requires careful thought about what is going on, but no calculations.

5.18   **Pumping Efficiency in Different Lasing Regimes**

a. Use a **Flow Rate** screen for 2→1 lasing. Start from default values, then set log A_{21} to about -8.0 and set log pump power to about 7.5. Record all three populations. Study the flow rates and record the atoms/sec between each set of levels for both spontaneous and stimulated emission. Go to **Plots** and plot **Output Intensity** versus **Mirror Transmission**. Record the value of mirror transmission at which there is no more coherent output.

b. Without changing any parameters, go to the **Flow Rate** screen for 1→0 lasing. Change A_{21} back to -6.0, and set A_{10} to around -8.0, leaving everything else the same. Record all three populations, and record the atoms/s of both types between each pair of levels. Plot **Output Intensity** versus **Mirror Transmission**, and make a sketch of that plot to turn in.

c. The radiation field between levels 0 and 2 has been enhanced by the same factor in parts a and b. What other factors were kept the same between parts a and b?

d. Study your results from parts a and b, especially the flow rate information. Write out a qualitative explanation of the differences in behavior
between the two lasing regimes. The differences must in some way
be due to the different A_{21} and A_{10} values. Try to qualitatively explain
how these create flow rate differences which lead to the results in
parts a and b.

5.19   **Atomic Flow Rates Near Thermal Equilibrium**

a. Show that the following approximations apply when n_0, n_1, and n_2
are fairly near their thermal equilibrium values.

$$F_{20} \approx A_{20}(n_2 - n_0 e^{-E_{20}/kT}), \qquad (5.35)$$

$$F_{21} \approx A_{21}(n_2 - n_1 e^{-E_{21}/kT}), \qquad (5.36)$$

and

$$F_{10} \approx A_{10}(n_1 - n_0 e^{-E_{10}/kT}). \qquad (5.37)$$

Be sure to write down your starting equations. Make it clear why
certain terms are neglected with respect to others.

b. Rewrite Eqs. 5.35–5.37 for pump power P turned on between level 0
and level 2, and populations still near their thermal equilibrium values.

c. If A_{21} is made small compared to A_{20} and A_{10}, discuss in qualitative
terms what happens to the flows of atoms between levels in part b.
As pumping begins, which levels are affected? Which populations are
large changes in the starting population and which changes are small
changes in the starting population? Include changes in all three levels
in your discussion.

5.20   **Pump Power and Population Inversion**
The "threshold of population inversion" occurs for 2→1 lasing where
$n_2 = n_1$.

a. Simulate population inversion for 2→1 lasing, using the default parameters. Increase pump power, observing that the n_1 remains always
greater than n_2. Then reduce log A_{21} somewhat to around −7 or so.
Again increase the pump power until the $n_2 = n_1$, and record the
value of log pump power. Write down your observations about the
changes in n_0, n_1, and n_2 while pump power was increased.

b. Explain why the threshold of population inversion is not the "threshold
of lasing."

c. Derive an approximate expression for the pump power P when $n_1 =
n_2$ for $A_{20} = A_{10}$, and $A_{21} << A_{20}$ (when $T = 300$ K). Hint: Review
the ideas in the previous problem.

d. Show that your derived result in part c is in good agreement with the
simulation results from part a.

5.21   **Cavity Stability and Cavity Shape**

a. Bring up the cavity simulation. Press **InsertRay** and observe that a
single ray is injected into the cavity and bounces back and forth. The

bouncing stops at 100 bounces, or when the ray angle to the axis gets too large. You should observe a stable bouncing pattern (i.e., the rays do not go out of bounds).

b. There is a simple algebraic stability condition[3] for the cavity:

$$0 < g = (1 - \frac{D}{R_l})(1 - \frac{D}{R_r}) < 1, \qquad (5.38)$$

where D is the distance between mirror vertices, and R_l and R_r are the radii of curvature of the mirrors. R is taken positive if the mirror is concave as seen from the middle of the cavity, and negative if R is convex. Read the values of D, R_l and R_r from the screen and calculate g. Report these values and whether the cavity is stable or unstable.

c. Repeat parts a and b using a *negative* value for R_l, the radius of the left-hand mirror. Find and report a value of R_l where the cavity is unstable, and one value of R_l where the cavity is stable.

5.22 Determining the Rayleigh Range z_o

Use the default simulation to launch (insert) a ray in the cavity. After the rays have stopped bouncing, select the beam width mode.

a. Measure the "raw" (z') value of the beam waist and the z'-value of each mirror, then subtract z'_{waist} to obtain z-values for each mirror.

b. Use the z- and R-values for the right-hand mirror (both>0) to calculate z_o. Then use the z- and R-values of the left-hand mirror (both<0) to determine z_o. These two values should agree to within a few percent.

c. Determine the Rayleigh range z_o directly from beam width measurements and compare to your calculated values from parts a and b.

5.23 Checking Beam Width Behavior

Use the default simulation to launch (insert) a ray in the cavity. When the rays are done bouncing, select **BmWidth** and determine z_o as in part c of the previous exercise.

a. Measure the beam waist w (it is half the length of the adjustable vertical bar), and the "raw" $z(z')$-value at the waist.

b. Measure the beam width and z'-value at the point where the beam strikes the right-hand mirror.

c. Use your value of z_o and Eq. 5.32 to calculate the beam width at the right-hand mirror, and compare to your measured value.

d. Repeat parts b and c at the left-hand mirror.

5.24 Rayleigh Range z_o at Small Launch Angle

Use the default simulation to launch (insert) a ray in the cavity. Select a launch angle of about 0.04 radians, and set the y-magnification for 10$\times$. This scheme should permit a more precise determination of the beam waist location, since the curvature of the beam is more evident.

a. Measure the z'-value of the beam waist and the z'-value of each mirror, then subtract z'_{waist} to obtain z-values for each mirror.

b. Use the z- and R-values for the right-hand mirror (both>0) to calculate z_o. Then use the z- and R-values of the left-hand mirror (both<0) to determine z_o. These two values should agree more closely than those of the earlier exercise Determining Gaussian Beam Parameter z_o.

5.25 **Create a New Stable Cavity and Measure Gaussian Beam Parameters**

Change the mirror radii (under **Set Cavity Parameters**) and launch rays in the cavity until you have found a new stable cavity. Then determine z_o as in the preceding exercises, explaining and showing your work. Using your z_o-value, check the width of the Gaussian beam at three or more points.

References

1. Planck, M. *Verh. d. D. Phys. Ges.* **2**:237, 1900.

2. Einstein, A. Phys. Z. **18**:121, 1917.

3. O'Shea, D.C., Callen, W.R., Rhodes, W.T. *Introduction to Lasers and Their Applications.* Reading, MA: Addison-Wesley, 1977, p. 74.

4. Saleh, B.E.A., Teich, M.C. *Fundamentals of Photonics.* New York: John Wiley & Sons, 1991, Chapters 3 and 9.

5. Schrodinger, E. *Statistical Thermodynamics.* Cambridge University Press: New York, 1960, Chapter II.

6. Andrews, D.G.H., Tilley, D.R. A computer model of laser action in the teaching of computational physics. American Journal of Physics, **59**:536–541, 1991.

7. Dohner, H.-J., Elsasser, W. Analysis of a four-level laser system: Investigations of the output power characteristics of a He-Ne laser. American Journal of Physics, **59**:327–330, 1991.

8. Stanek, F., Tobin, R.G., Foiles, C.L. Stabilization of a multimode He-Ne laser: A vivid demonstration of thermal feedback. American Journal of Physics, **59**:932–934, 1993.

6

Nuclear Properties and Decays

Michael J. Moloney

6.1 Introduction

This simulation consists of three modules. The first module on nuclear properties is in program NUCLEAR.PAS. The second and third modules on nuclear counting and decay rates are in program DECAYS.PAS.

6.2 Module on Nuclear Properties

6.2.1 Nuclear Constituents and Nuclear Binding Energy

The atomic nucleus is considered to be made up of protons and neutrons. The number of protons is called the atomic number, Z. The number of neutrons is simply called the neutron number, N. The atomic mass number, A, is the sum of the protons and neutrons, collectively known as nucleons ($A = Z + N$).

The mass of each nucleus is less than the sum of the masses of its component neutrons and protons. The fact that a nucleus has less mass-energy than its components can be understood in terms of its binding energy, the amount of energy required to split the nucleus into its isolated protons and neutrons. The simplest nuclide is $^{1}_{1}\text{H}$, which has zero binding energy because it is already an isolated proton. The next simplest nuclide is $^{2}_{1}\text{H}$ (deuterium), whose binding energy is found by summing the masses of a proton and neutron, and subtracting the nuclear mass of deuterium, then multiplying by c^2, where c is the speed of light in vacuum. Deuterium is probably the only case in which we could supply energy (say, in the form of a gamma ray) and split this nuclide into its separate, isolated protons and neutrons. Binding energy for all nuclides is calculated in the same way; add the masses of all protons and neutrons, and subtract the nuclear mass of the intact nuclide. This gives a "mass excess" which (multiplied by c^2) gives the binding energy. The greater the binding energy per nucleon, the greater the stability against wholesale breakup of the nucleus.

6.2.2   Semi-Empirical Mass Formula

Nuclei are found to have fairly constant mass density, so the radius of a roughly spherical nucleus is proportional to the cube root of its mass (or atomic mass number, A):

$$r = r_o A^{1/3}, \tag{6.1}$$

where r_o is around 1.3×10^{-15} m.

The semi-empirical mass formula (SEMF) was first developed by von Weiszacker in 1935, based on similarities between a nucleus and a drop of liquid, but including systematic features of nuclear masses. Many forms of the formula exist. The one given below was used by Green in the 1950s, and is given in Leighton's *Principles of Modern Physics* text[1]: (for binding energy, abbreviated B.E.)

$$\text{B.E.} = c_1 A + c_2 A^{2/3} + c_3 Z(Z-1)A^{-1/3} + c_4(N-Z)^2/A + c_5/A, \tag{6.2}$$

where c_1 is positive, while c_2, c_3, and c_4 are negative. The sign of c_5 is positive if the nucleus is even-even (even-Z, even-N), negative if the nucleus is odd-odd, and zero otherwise. The c_1 term represents the attractive effect of the nuclear force. c_2 represents a negative effect of the surface area; surface nucleons are not as strongly bound as interior nucleons. c_3 represents the tendency of proton repulsion to disrupt the nucleus, and c_4 represents the reduction in stability due to an excess of either protons or neutrons. c_5 represents the fact that a single nucleon in a given state is not as strongly bound as are two (spin-paired) nucleons of the same type in this same state.

For A = constant, Eq. 6.2 shows that B.E. is a quadratic function of Z. When A is odd, nuclei are even-odd or odd-even and c_5 is zero. A plot of B.E. vs. Z for odd-A should therefore produce a single parabola. But when A is even, nuclei are either even-even (and $c_5 > 0$) or odd-odd (and $c_5 < 0$). So when one plots B.E. vs. Z for even A, there are two parabolas, with the one for even-even nuclei lying above the one for odd-odd nuclei. This module contains a parabola "fit" in the simulation so that plots like this can be examined, to see how close to parabolic they are.

In chemistry, we are familiar with the special stability of the noble gases. Their stability is due to closed shells of electrons. Nuclei also display special stability associated with closed shells of protons or neutrons. This behavior is seen at the "magic" numbers of 2, 8, 20, 50, 82, and 126. A nucleus with a magic number of protons and a magic number of neutrons is said to be "doubly magic," and should be especially stable. Examples of doubly magic nuclei include $^4_2$He and $^{208}_{82}$Pb.

6.2.3   Nuclear Decays

There are three types of nuclear radiations—alpha, beta, and gamma.

Alpha particles are $^4_2$He nuclei emitted by very large nuclides, mainly those of lead or heavier elements. Proton repulsion energy in nuclei depends on the square of the atomic number Z. It is relatively small in light nuclei, but is a major factor in heavy nuclei, where it becomes energetically feasible to eject a doubly charged alpha particle.

All alpha particles emitted in the decay of a specific nuclide have the same energy, but when alpha decays from all nuclides are surveyed, it is found that most alpha particles have energies in the range of 4 to 8 MeV.

Alpha emission involves the loss of four mass units, so there are four decay "series" which run roughly from uranium down to lead. The half-lives of alpha decays in this series range from billions of years to tiny fractions of a second.

Quantum mechanics had a great early success by explaining alpha particle half-life and energy behavior, using a model where the alpha particle was taken to be trapped in a potential well inside the nucleus. The theory was able to account for the enormous range of half-lives because of exponentially varying likelihood of penetrating the barrier of the potential well, and was also consistent with the range of observed energies.

The results[3] of a quantum-mechanical calculation for alpha decay look like

$$\log_{10} \lambda = C + 1.28 Z^{1/2} R^{1/2} - 1.71 Z E^{-1/2}, \tag{6.3}$$

where C is a constant, λ is the decay constant for the decay (the probability of decay per nucleus per unit time), Z is the atomic number of the residual nucleus (after the decay), R is the radius of the nucleus (in units of 10^{-15} m), and E is the energy of the alpha particle in MeV. This equation is the basis for two of the plots which are available in this module.

Beta particles may be positive (positrons) or negative (electrons). Beta-minus particles come from nuclei which are unstable due to too many neutrons. Beta-plus particles come from nuclei which are unstable due to too many protons. Either type of beta decay emits both a beta particle and a neutrino or antineutrino (one is the antiparticle of the other; both are light particles which are extremely hard to detect). Because two particles (the beta and neutrino) are "competing" for the available energy in the decay, some betas have more energy and some have less.

Gamma rays are chunks (or quanta) of electromagnetic energy like photons of visible light. Gammas are emitted from nuclei which are in an excited state; emitting the quantum of electromagnetic energy permits the nucleus to reach lower energy states, and eventually its ground state.

6.2.4 Nuclear Reactions

Nuclear reactions are characterized by a shorthand indicating the incident particle and final particle. A reaction in which a neutron came in, struck some nucleus, and a proton was ejected would be called an (n,p) reaction. An explicit example of an (n,p) reaction is

$$n + {}^{14}_{7}N \rightarrow {}^{14}_{6}C + p. \tag{6.4}$$

This reaction is one which forms ^{14}C in the Earth's upper atmosphere.

The user may select from among six incident particles, and the same six ejected particles: gamma ray, proton, neutron, deuteron, triton, and alpha particle.

One column of reaction choices appears at a time, with a common incident particle. The → **More** (and **More** ←) choices take the user to a column of entries for the next (or previous) incident particle. The lists are presented in order of increasing mass of the particles (γ, p, n, d, T, α).

The Q-value of a reaction is the energy released in the reaction. Negative Q-values mean that the reaction requires energy to be supplied by the incident particles. There will be a positive reaction Q if the total rest masses of the final state particles is less than the total rest mass of the initial state particles.

For all nuclear decays and nuclear reactions, the Q-value is printed on the upper right-hand side of the screen under the information identifying which nuclide has been selected by the "active" edit cursor. If a Q-value cannot be calculated due to a mass missing from the database, a value of 99.9 MeV is assigned.

The case of an incident gamma ray involves the same energetics as particle emission. A (γ, α) reaction is physically different from alpha particle emissions, but shares the same rest masses and energy requirements. An (n,γ) reaction means that a neutron comes in, strikes the target nucleus, and only a gamma ray is emitted. This is "absorption" of a neutron; no distinct nuclear particle emerges.

6.2.5   Delayed Neutrons

Nuclear fission is the breaking up of heavy elements into fission fragments, which are medium-mass nuclides. Because the neutron/proton ratio increases throughout the periodic table, fission fragments have too many neutrons to be stable. Beta-minus emission is the main feature of their decays, and fission fragments can occasionally emit a neutron after a beta-minus decay. Such neutrons are known as delayed neutrons.

Only two nuclides in the database show up as being capable of pure neutron emission from the ground state. One has an improbable Q (probably an incorrect mass from the data table). The other can be regarded as possibly a genuine neutron emitter. These two nuclei aside, no two isotopes of the same element are separated by as much as or more than the mass of a neutron. Yet, a dozen or so delayed neutron emitters are known to exist among the fission fragments.

These delayed neutron emitters can emit neutrons only if they are formed after beta-minus decay in a sufficiently excited state to have a positive Q for neutron emission.

The module offers a late neutron option which will graph all nuclides and show which are capable of delayed neutron emission. The nuclide shown is not the one which will actually emit the delayed neutron, but rather the "mother" nuclide which decays into a daughter which is the potential emitter of delayed neutrons.

6.2.6   Module Database

This module utilizes a nuclear database consisting of masses, lifetimes, and primary decay modes. It is rather comprehensive (about 1,900 nuclides) and generally quite accurate, typed in from the text by Blatt.[2] It has been extensively tested and rechecked.

The module requires three files: NUCLEAR.DAT, NUCLISTS.DAT, and NUCELEM.DAT. This last file contains only the two-letter abbreviations of the elements. The first two are created by a program READDATA.PAS, which reads and parses an ASCII nuclear data file called BLATT.DAT, containing data from Blatt's Appendix B.

6.2.7 Running the Program

Using Menus

The **Pick Plot** item on the top menu is used to select the type of plot to be done. Within each type of plot there are subchoices for that plot, and specific instructions to create the graph when the user is ready.

Choice of parameters is mainly from bar menus. Arrow keys move among entries on a bar menu, and selection is done via the **Enter** key, or with a mouse click. When a number of choices are available on a given bar, one repeatedly selects that bar until the correct choice is shown. The program has some checks on selections, to avoid conflicts of the following type: Plot y-axis = B.E./A, x-axis = Z, at constant Z. Because of these checks, a wrong plot can occasionally come up since in the process of choosing the parameters, one may have changed due to a conflict, and the user did not notice it. In this case, one just goes back to the **Do Plot** menu and corrects the situation.

Plots Involving the Binding Energy

B.E./A (a quantity directly connected to nuclear stability) may be plotted versus Z, N, or A. The plot may be done for stable nuclides only, or for all nuclides. The plot may be unrestricted, or may be made such that any value of N, Z, or A can be held constant.

Mass Formula Plots and "Magic Numbers"

Plots are available for the semi-empirical mass formula (SEMF), including one with B.E./A on the y-axis. This is especially interesting since you can view the B.E./A curve itself. The colors tell you whether the B.E./A calculated from the SEMF is higher or lower than the experimental B.E./A. A blue color says that the SEMF B.E./A is too low, and means that nuclei in this region are more stable than predicted from overall "liquid" nuclear characteristics. This blue color can be an indicator of a region of special stability, as one finds in the "closed shells" in the periodic table of the elements. For nuclei, there are said to be "magic numbers" of either neutrons or protons where one finds "closed shells" inside the nucleus. Even though nuclear shell structure is beyond the scope of a modern physics course, one can look for evidence for "magic numbers," and this plot is one way to do it.

Nuclear Decays and Nuclear Reactions

To learn about nuclear decay energetics, and the energetics of nuclear reactions, one goes to **Pick Plot** on the top menu and selects **Nuclear Decays/Reactions**.

This brings up an input screen from which you can choose what kind of process (decay or reaction) will be plotted, and how. Initially, the choices are for decays (particles emitted are beta-plus, beta-minus, proton, neutron, alpha, and late [delayed] neutrons). There also is a full matrix of reactions (a,b), where a is the incoming particle and b is the emitted particle. The set of incoming or outgoing particles is the same: γ, p, n, d, α, and T.

One might select a (p,γ) reaction (by clicking on that item, or using the arrow keys to reach that entry and then pressing **Enter**). Then doing the plot will show all possible (p,γ) reactions. The color is different depending on whether the Q-value exceeds the one selected (by **Minimum particle energy**), but one can examine the Q-value of any nuclide by bringing the cursor to that location on the graph. The upper right-hand part of the graph identifies the nuclide selected and the Q-value for its reaction (or decay, if a decay was selected).

For decays, one may plot those nuclei which have enough energy to decay into others by emitting an alpha, beta-minus, or beta-plus particle. Decay energy may be specified so that, for example, one could plot all alpha emitters with energy greater than 4.2 MeV. Nuclides in this plot are not necessarily experimentally observed alpha emitters; they are those whose alpha decay energy is possible above the minimum energy. In this sense, the plot is theoretical. (Some decays will not show up due to database limitations. If a "daughter" nuclide is not in the database, the "parent" nuclide of a given decay will not be plotted.)

Plots of Alpha Decay Lifetime and Energy

Three plots are available which relate alpha particle decay lifetimes and energies. All three have the natural logarithm of the alpha particle half-life plotted on the y-axis. The x-axis of the first plot shows alpha particle energy E, that of the second plot shows reciprocal square root of energy ($E^{-1/2}$), and the x-axis of the third plot shows $ZE^{-1/2}$, where Z is the atomic mass of the residual nucleus (after the decay). The first plot is suggestive of a simple functional relationship, and the last two are motivated by the quantum-mechanical calculation, expressed by Eq. 6.3. The last two plots are organized by increasing Z, so that one may move the edit cursors to display only those alpha decays from a given Z-value. The cursor identifier always refers to the initial nucleus, of course, before the decay.

In all of these plots the decays which lie in the "series" from uranium down to lead are plotted in a different color. This helps to distinguish behavior of the scattered alpha emitters below lead, from those in the main series, and from the elements above lead in the periodic table.

6.3   Module on Nuclear Counting

6.3.1   Introduction

Simulated Experiment

This module simulates an experiment using a Geiger-Muller tube (GMT) repeatedly to measure the decays from a long-lived radioactive source. Each measurement lasts for the same fixed counting time. The GMT registers a count via an avalanche ionization process which produces a large voltage pulse, and leaves many gaseous ions still in the chamber. After recording a pulse, the GMT must "recover" its original state by rebuilding the original potential on its central electrode. For a time called the "dead time" or the "resolving time" of the GMT, it is not able to respond to another ionizing event, because it is recovering to its original state.

Measuring Experimental Geiger Tube Dead Times

One can readily measure the dead time of a Geiger tube using an oscilloscope, a GMT, and a microcurie source of beta or gamma radiation. The radioactive source should be carefully set a few millimeters from the GMT end window. (Be sure to put adequate shielding between you and source, especially if it's a source of gamma rays.) This will send GMT counts to the y-input of an oscilloscope so that each count can trigger a fresh sweep. The dead time shows up as a blank space on the left-hand side of the screen, while lots of pulses "dance around" on the screen to the right of the blank space. The size of the blank space is approximately equal to the GMT dead time, since no counts occur in there after the sweep is triggered.

6.3.2 Counting Statistics With and Without Dead Time

If we expected on the average to measure E counts in time T, and if there were no GMT dead time, the probability of measuring exactly n counts in time T would be the Poisson distribution:

$$P_n(E) = \frac{E^n e^{-E}}{n!}.$$ (6.5)

But in the presence of dead time, all counts which occur during the recovery time of the GMT are not recorded. The tube is considered "dead" during the recovery time following the recorded GMT event.

   If the GMT (whose dead time is D for each count) records exactly G counts during time T, then it was "dead" for a time DG, and "live" for a time $T - DG$. The "true" count rate may be estimated as the number of counts divided by the "live" time: $G/(T - DG)$. Setting this equal to E/T, the expected ideal count rate, we get

$$\frac{E}{T} = \frac{G/T}{(1 - D(G/T))}.$$ (6.6)

Rearranging this gives

$$1/E = 1/G - D/T.$$ (6.7)

6.3.3 Simple Limiting Cases for *E, G,* and *D*

Equation 6.7 says that E and G will be very close when D is a small fraction of the time T during which we are counting. That is, if $D/T << 1$, G is very nearly the true count E. If D/T approaches or exceeds 1, G approaches zero. (One normally would not choose a counting interval T smaller than the GMT dead time D.)

If E becomes very large, then G approaches T/D. This means that the tube counts continuously. E is so large that every time a "dead time" expires there is an event waiting to be measured. The distribution becomes really squashed here; there is almost no variation in counts—each measurement is right around T/D.

With non-zero dead time, high count values during time T tend not to be as high because of more dead time and less "live" time, while low count values during time T tend not to be as low because of more "live" time to count. Thus, when we look at histograms (plots of how many times a given count was recorded vs. the count) in the presence of dead time, we expect to see fewer extreme values; the distribution will be somewhat compressed toward the expected value at the center.

6.3.4   Walk-Through Tutorial for Nuclear Counting

There is a tutorial on nuclear counting, especially as it applies to dead time in Geiger-Muller tubes. This tutorial is selectable under **Tutorial** on the top menu. The tutorial is intended to clarify the notion of ideal counts (which would be registered on an ideal counter with no dead time), and GMT counts, which are those registered on a counter with finite dead time.

6.4   Nuclear Buildup and Decay Rate Module

6.4.1   Decay of Radioactive Nuclei

Alpha, beta, and gamma decays make up what is called "radioactivity." All three types are described by the same equation for the average number of survivors, N, as a function of time. In each case the probability of decay per unit time per nucleus (λ) is a constant. Multiplying λ (the decay constant) by the number of undecayed nuclei, N, gives the rate of decrease of the surviving nuclei, $-dN/dt$:

$$-\frac{dN}{dt} = \lambda N. \tag{6.8}$$

This equation is readily solved to yield

$$N(t) = N_o e^{-\lambda t}. \tag{6.9}$$

This equation for $N(t)$ is a good description of the average behavior of a large number of nuclei. When $N(t)$ is returned to the decay rate equation, we have the decay rate as an exponentially decreasing function of time:

$$-\frac{dN}{dt} = \lambda N_o e^{-\lambda t}, \tag{6.10}$$

where N_o is the number of nuclei at $t = 0$. People often refer to the "half-life," $t_{\frac{1}{2}}$ of a nuclide, rather than its decay constant, λ . The two are readily related by setting $N(t_{\frac{1}{2}}) = N_o/2$, resulting in

$$t_{\frac{1}{2}} = \frac{\ln 2}{\lambda}. \tag{6.11}$$

6.4.2 "Activating" Material With Neutrons

Since neutrons are electrically neutral, they are not repelled by protons in the nucleus, and may enter a nucleus if they get near it. A stable nucleus entered by a neutron may become radioactive ("activated") and then decay, often via negative beta decay.

Many colleges have "neutron howitzers," distributed by the predecessor of the Department of Energy (i.e., the Atomic Energy Commission) in the 1950s and 1960s. These are plexiglass, water-filled cylinders with a central shaft containing a fist-sized plutonium-beryllium slug. This slug generates the neutrons, which are slowed down by collisions in the water. Materials placed near the center of these howitzers can be activated by absorbing neutrons, and their radioactivity measured.

In this simulation, turning on the flux is comparable to putting some material in the howitzer, and turning off the flux is comparable to removing the material from the howitzer to have its decays counted. One can place objects at different distances from the center of the howitzer to vary the neutron flux they receive, but usually one uses a region of fairly constant flux within 6 inches of the central plastic shaft.

6.4.3 Using the Decay and Growth Rates Module

Types of Runs

There are two types of runs which can be done in this module.

- **1. Series**: Species 1 decays into species 2, which decays into species 3.
  The user controls of the half-lives of the three species. Plots are available which show population of each species vs. time, decay rate of each species vs. time, and natural log of decay rate vs. time. Graph editing, line fitting, and parabola fitting (see below) may all be used.

- **2. Activation**: Two nuclides are activated by neutron flux, which can be turned on and off by the user.
  There are two species being activated by neutrons. Each has an adjustable capture cross section, and half-life of the activated nuclide. Population and decay graphs are available as in item 1, as well as the indicated graph manipulation features.

6.5 *Features of the Program*

6.5.1 Graph Editing

All graphs are initially shown in **Edit** mode. Two cursors are available in this mode; the "active" cursor sits on a data point whose properties are shown in detail in the upper right-hand part of the screen. Moving the active cursor to different data points lets one see any detailed value from the graph. (There is also a "passive" cursor, but the properties of its data point are not shown.)

To look at a subset of all the graph data, one "zooms in" by positioning the two cursors, and then invoking the zoom function, which replots all the data between the cursors.

Keys are available for all cursor manipulation and for zooming in and out (expanding the graph back to the full data set), but mouse control may also be used. The mouse can "drag" an edit cursor from one data point to another. A sound will indicate that the cursor has been selected, but the cursor will not appear to follow the mouse. When the mouse button is released within 5 screen pixels of a data point, the cursor will appear around that data point. If no data point is within 5 pixels of a data point, a sound will indicate that the drag was unsuccessful. The distance at which the mouse can "grab" a cursor is 20 pixels; this means that for short moves of the cursor one may wish to point at the desired nucleus, click and let go. This will grab and move the cursor.

It should be noted that the cursors are not permitted to cross; the left cursor must stay to the left of the right cursor.

"Fitting" modes are described next; the user may switch between editing and fitting graphs using keystrokes or via the mouse.

6.5.2   Graph "Line and Parabola Fitting"

All graphs may be "fitted" with a straight line, by entering "line" mode. Two cursors are presented, which the user may move around. The equation of the line is available on demand, but it is the responsibility of the user to judge whether the equation makes sense. This function is thus not a statistical operation; it merely lets the user exercise judgment.

In a similar fashion, there is "parabola" mode for presenting a parabola on any graph. There are three cursors to define the parabola, and its equation is on call.

6.6   Exercises

6.6.1   Exercises for the Nuclear Properties Module

6.1   **Recommended Walk-Through Tutorial**
It is highly recommended that you first run the **Tutorial Walk-Through**, which is available under **File** on the top menu. This walks through the use of edit cursors, zooming in and zooming out in various plots, and fitting via lines and parabolas. An investment of five or ten minutes doing this tutorial will be well repaid.

6.2   **Ratio of Neutrons to Protons Using Line Fits**
a. Under **Mass Formula**, plot N versus Z for stable nuclides, leaving other choices alone. Zoom in on the region from $Z = 1$ to $Z = 30$. (Select the right cursor [press **Tab**], move it left, then press **End**). Fit the lower part of this region with a straight line (press **L** to enter line mode). Look at the equation of this line, which should have a slope a

little greater than 1. This just says that for the lighter elements stable nuclides have about as many protons as neutrons.

b. Return to **Edit** mode (press **E** or use a hot key) and zoom back (**Home** or hot key) to the full graph. Try fitting a straight line to the latter two-thirds or so of the graph, after the lightest elements. Write down the equation of this line. How many neutrons are being added per proton in stable nuclides over this range?

c. What is the ratio of neutrons/protons in the heaviest nuclides, roughly? (Hint: Select the right cursor, which will show the parameters of the nuclide in the upper right-hand corner.) Does the answer to part b contradict the answer you just gave, or are they consistent? Briefly explain.

6.3 **Most Stable Element, and Next-Most-Stable Element**
Plot binding energy per nucleon (B.E./A) versus Z for all nuclides.

a. Check on the claim that iron is the most stable element, citing the evidence in favor of the claim.

b. Figure out the main challenger to iron for the title of "most stable element" and cite the evidence in favor of the challenger. Hint: Zoom in on the region around the peak in the B.E./A graph.

6.4 **Mass Formula as a Fit to All Nuclides**
Select **Mass Formula** plot, N vs. Z, for all nuclides. Note the many nuclides which are in red or blue, meaning that their experimental mass does not fall within an error margin of 0.01 MeV of the **Mass Formula** value. Go back to the input screen by selecting **N-New Data** and gradually increase the error margin from 0.01 MeV until 90% or so of the nuclides in the database are in green. Record the value of the error margin. What region remains with parts of it outside the error margin? What is there about the mass formula (and the liquid drop idea on which it's based) which would suggest that the mass formula might not be a big success in this region?

6.5 **Magic Numbers Based on Stable Nuclides**
The idea here is to look for magic numbers of protons and neutrons where there are a lot of stable nuclides for a given N or Z.

a. Plot mass versus N for stable nuclides only. Look at values of N between 0 and 60. What values of N have the most stable nuclides? List the best six N values and how many stable nuclides each has.

b. Plot mass versus Z for stable nuclides only. Look at values of Z between 0 and 60. What values of Z have the most stable nuclides? List the best six Z values and how many stable nuclides each has.

c. List your top four candidates for magic numbers, based on numbers of stable isotopes for N and Z.

d. The magic numbers between 0 and 60 are supposed to be 2, 8, 20, and 50. What reason might there be for not all of these making your list?

6.6 **Magic Numbers Based on Binding Energy**
a. Pick the semi-empirical mass formula, and select a plot of binding energy per nucleon versus N, for all nuclides. Leave the error margin at 0.01 MeV. The blue areas on the plot are those where the mass formula

is too low (that is, the experimental B.E./A is higher than the general behavior as predicted by the mass formula). Assume that regions of higher stability may have to do with closing shells of neutrons in the nucleus, so the blue areas may be around magic numbers. Decide on a set of six candidates for magic numbers for neutrons, based on the blue areas, and regions of higher B.E./A.

b. Repeat part a except to plot B.E./A versus Z. Decide on a set of six candidates for magic numbers for Z, based on the same arguments (blue areas, greater B.E./A) as in part a.

c. Did you come up with exactly 82 and 126 for additional magic numbers? How does the evidence look to you, just based on this method?

6.7 **Nuclides Which Emit Alpha Particles**

a. Plot the log of half-life vs. alpha energy. This plots all the nuclides listed in the database as alpha emitters, and it shows their energies. From the plot, make a list of how many alpha emitters have an energy between 0 and 2 MeV, how many in the range 2–4, 4–6, 6–8, and 8–10 MeV. (Hint: Use the plot sorted by alpha energy. Moving the edit cursor across the data can be done in steps of 1, 10, or 100.)

b. Do a plot of potential alpha emitters all across the periodic table, setting the decay energy to 10 MeV. (Under **Pick Plot**, select alpha, beta stability. Under **Do Plot** set for alpha decay, 10 MeV.) This will tell you how many nuclides could potentially decay, emitting an alpha particle of 10 MeV or more. Repeat with an energy of 8 MeV, and get a count of potential alpha emitters between 8 and 10 MeV. (Zoom in to do the count.)

c. Repeat for alpha decays of zero energy. Make an approximate count of the nuclides which could potentially emit alpha particles with energies greater than 0 MeV. This should be a lot larger than the number found in part a. Why are so few of the potential alpha emitters actually seen in nature?

6.8 **Parabola Fits of B.E./A vs. Z at Constant Atomic Mass Number (A)**

a. Select a graph of B.E./A versus Z, at constant A. Start with a value of $A = 133$. Fit this set of points with a parabola. Note that by pressing the **End** key you can get the equation of the parabola.

b. Study the semi-empirical mass formula (Eq. 6.2) and explain why there will only be one parabola for constant A when A is odd, while there will be two parabolas when A is even.

c. Try a few even numbers for $A = $ constant (for example, even numbers between 130 and 150). For an even A there should be two parabolas. Even Z-values fit on one (even-even nuclides) and odd Z-values fit on the other (odd-odd nuclides). Try to find an even A where you can imagine two parabolas which should have the same shape, but one merely shifted higher up than the other. ($A = 148$ is a reasonable choice to start with.) Press **P** to get a parabola fit, then move the cursors to fit the upper parabola. (Don't forget that each parabola involves every other mass in the plot. One parabola for even-Z, and one for odd-Z.) Obtain the equation of the parabola, and write it down.

    d. How far below this parabola are the "other" masses (the ones you did not fit), on the average?

    e. Review the discussion of the semi-empirical mass formula, and calculate the difference which should exist between the even-even parabola and the odd-odd parabola. To do this calculation, locate the value on the input screen. Explain how you made use of the c_5 value in calculating the difference between parabola values. Compare the difference you calculated with the results in part c.

6.9 Find Q-Values for Carbon Cycle Reactions

In the Sun, the following steps of the carbon cycle liberate energy and provide much of the energy output (ref. 2, p. 386):

$$^{1}\text{H} + {}^{12}\text{C} \rightarrow {}^{13}\text{N} + Q_1 \tag{6.12}$$

$$^{13}\text{N} \rightarrow {}^{13}\text{C} + e^+ + \nu + Q_2 \tag{6.13}$$

$$^{1}\text{H} + {}^{13}\text{C} \rightarrow {}^{14}\text{N} + Q_3 \tag{6.14}$$

$$^{1}\text{H} + {}^{14}\text{N} \rightarrow {}^{15}\text{O} + Q_4 \tag{6.15}$$

$$^{15}\text{O} \rightarrow {}^{15}\text{N} + e^+ + \nu + Q_5 \tag{6.16}$$

$$^{1}\text{H} + {}^{15}\text{N} \rightarrow {}^{12}\text{C} + {}^{4}\text{He} + Q_6 \tag{6.17}$$

    a. Go to the **Nuclear Decays/Reactions** section and from appropriate plots determine the Q-value of each reaction in the carbon cycle. Write down each reaction and its Q-value. Determine the total energy released by all steps in the cycle.

    b. Based on the overall effect having four hydrogen nuclei converted into two beta-plus particles and an alpha particle, determine the Q of the overall reaction from the masses.

6.10 Equation 6.3 for Alpha Decay

    a. Select a plot of natural log of alpha half-life versus $ZE^{-1/2}$ and obtain a slope for decays below lead, a slope for the "series" decays from uranium down to lead, and a slope for the transuranic elements.

    b. The slope in Eq. 6.3 is -1.71. Why are the slopes from part a positive?

    c. Show that the slope of -1.71 in Eq. 6.3 should become a slope of $+3.94$ in the graph of part a.

    d. Suggest an improvement over the strategy of part a for comparing to Eq. 6.3. Try out your strategy. Is it an improvement over part a?

6.11 Find The Problem in ^{28}Mg Beta-plus Decay

    a. Select a plot of beta-minus decays, and observe that the Q-value for ^{28}Mg is greater than 2 MeV. Write down the exact Q-value.

    b. Make an argument that beta+ decay of ^{28}Mg is unreasonable in view of its composition of protons and neutrons.

c. Presumably this result occurs because of an erroneous nuclear mass (in spite of the generally high quality of the data in ref. 2). Which specific masses might be in error?

d. Could any light be shed on this question by a plot of beta-minus decays? If so, try it and report your results.

e. Identify some possible symptoms of an erroneous mass (if a mass were too low or too high, where would that cause "odd" parameters to show up?) Find evidence in other plots (other than beta decay) of nuclear properties which points to the "defective" mass.

6.12 (n,α) Reactions

a. Go to the **Nuclear Decays/Reactions** menu and select (n,α) reactions. Do the plot and make a rough estimate by eyeball of the fraction of all nuclei which yield energy from an (n,α) reaction. Then estimate percentages of all nuclei which yield more than 1 MeV alpha particles in an (n,α) reaction.

b. Since there are plenty of neutrons in a nuclear reactor, and plenty of nuclei which will yield energy in a (n,α) reaction, why doesn't one see a lot of this in a nuclear reactor?

6.13 Equivalence of Beta-minus and (p,n) Reactions?

A nucleus will wind up with the same number of protons and neutrons by emitting a beta-minus particle as it will in a (p,n) reaction. Will this lead to exactly the same Q-value for both situations?

a. Go to the **Nuclear Decays/Reactions** menu and select beta-minus emission. Do the plot and select a nucleus which has a positive Q when it emits a beta-minus particle. Write down this nucleus and record the Q-value. Then go back and select (p,n) reactions, do a plot, and record the Q-value of that (p,n) reaction.

b. You should have found a discrepancy in part a. Try to determine why it is there. Write down your explanation, and make a calculation which supports your explanation.

6.14 Proton Emission versus (n,d) Reaction

Repeat the previous exercise, comparing the Q in proton emission (the Q's will be negative, but select some nucleus for comparison anyway) with Q in an (n,d) reaction, where a deuteron is emitted. Determine the cause of whatever discrepancy you find, and make a calculation which shows the exact amount of the discrepancy.

6.15 Delayed Neutron Emission

a. Go to the **Nuclear Decays/Reactions** menu and select (**late n**) reactions. This will show which nuclides are energetically capable of having a "delayed" neutron emitted after the original nuclide has first beta-minus decayed and the daughter nucleus *may* be capable of emitting a neutron. Select a nuclide from the middle of the periodic table, write it down, and also the amount of energy which might be released (the Q-value).

b. Select beta-minus decays and record the Q-value for that same nuclide for beta-minus decay.

c. Determine the identity of the daughter nucleus from the beta-minus decay in part b. Do a graph of pure neutron emission, and determine the Q-value for neutron emission from the daughter nuclide. (This Q will be negative, meaning that pure neutron emission is not possible from the ground state of the daughter nucleus.)

d. Connect all previous parts of this problem. Explain how neutron emission from the daughter might be possible, depending on the details of the original beta-minus decay.

6.16 **Energy Barrier to Neutron Absorption?**

From your experience with the previous exercise, think over the situation with spontaneous emission of a neutron from any nuclide. The reverse of this situation is a nucleus absorbing a neutron. Is there a nuclide in the database which will have a negative Q for absorbing a neutron? Without running the graph of (n,γ) reactions, decide if there are nuclides which will NOT emit energy after absorbing a neutron (they would need energy available before a neutron could be absorbed). Give your reasoning and your conclusion, understanding that you must do more than claim support from the (n,γ) graph.

6.17 **Neutron Excitation of Heavy Nuclides**

Fission is initiated by absorption of a neutron, which may or may not cause the resulting nucleus to break apart. This exercise asks you to explore which of the heavy nuclides have the greatest excitation (Q-value) after absorbing a neutron.

a. Go to the **Nuclear Decays/Reactions** menu and select (n,γ) reactions. Be sure you are plotting N on the y-axis and Z on the x-axis. Set a minimum energy of 4 MeV, so that all nuclides receiving more than 4 MeV will be highlighted. Then change the Q "barrier" to 5 MeV, look at the graph, set the barrier to 6 MeV, and look at that graph. Record the Q-values of ^{238}U, which is most of naturally occurring uranium (and not very fissionable if neutrons have very little energy), and of ^{235}U, which is scarce (but is definitely fissionable with neutrons of low energy).

b. Can you see some systematic behavior of nuclides above and below $Q = 6$ MeV? Try to identify the pattern in terms of neutron number, N, and proton number, Z.

c. ^{238}U is reasonably fissionable, with "fast" neutrons. Based on the results of part a, make an educated guess about how much kinetic energy neutrons would need in order for ^{238}U to have a fair chance of fissioning.

6.6.2 Exercises for the Nuclear Counting Module

6.18 **First Run With Default Values**

a. Do a run without changing the initial parameters, and let it go to completion at $t = 1000$ seconds. Select **Poisson** and move the curve until it fits over the **Ideal** (no dead time) bar graph. (You can move the curve by clicking the mouse near the peak and then "dragging," or by using the left and right arrow keys.) You want the area of the bars above the curve to be about that of the "missing" area of bars below the curve. Record the value of E when the curve fits as well as it can. Now move the curve to get a best "fit" to the **GMT** curve (including dead time effects). Record the value of E.

b.   i. Was the fit to the **Ideal** curve (select one)—very close, close, fair, not close, not close at all?

ii. Repeat step i for the fit to the **GMT** curve.

c. The value of E should definitely be higher for the **Ideal** curve than that for the **GMT** curve. Is it?

d. The E-value for the fit to the **Ideal** curve should be 25.0 counts/s because this is what is ideally expected. (In order to plot both **Ideal** and **GMT** curves, each is shifted a little bit. A count of 25 by itself would be a bar of width 1.0 centered on 25. When **Ideal** and **GMT** are plotted together an **Ideal** count of 25 is plotted as a bar of width 0.5, centered at 25.25, while a **GMT** count of 25 is plotted as a bar of width 0.5 centered at 24.75). How does your **Ideal** E-value compare with the expected E now that you know all **Ideal** bars are shifted up 0.25 and all **GMT** bars are shifted down 0.25?

e. Use the **Ideal** and **GMT** values of E to examine Eq. 6.7. How close is the left-hand side to the right-hand side of this equation?

6.19   **The Limiting Case of Large E**
a. Change the ideal count rate to 1500/s from the default value of 250/s under **Set Parameters**. Change the number of samples to 4000 from the default value of 10,000; do a run and let it go to completion at $t = 400$ s.

i. Do the same curve fits as in Exercise 6.18. How close are the curves to the bar graphs? Can you see any "shrinking" in the distribution of bars for the **GMT** simulation?

ii. How well does the relation $1/E = 1/G - D/T$ hold up here? (Calculate the left-hand side and right-hand side, and compare them.)

b. Change the ideal count rate to 3000/s from 1500/s and let it go to completion at $t = 400$ s.

i. Is the distribution of **Ideal** bars close to a poisson curve?

ii. Ditto for the distribution of **GMT** bars.

iii. Pretend you don't know the **Ideal** count rate, but you do know the experimental one, and the dead time (just as you might during a laboratory experiment). Calculate the ideal count rate and compare it to the value of 3000/s.

6.20   **Limiting Case of Small D**
a. Use the default parameter settings, except for setting the dead time value to a value of 50 μs, and do a run to completion at $t = 1000$ s.

b. How close are the **Ideal** and **GMT** bar graph curves? (Remember that the **GMT** curve is artificially shifted 0.25 down, and the **Ideal** curve is artificially shifted 0.25 up so they can both be plotted on the same graph. See Exercise 6.18d.)

6.21 **Limiting Case of $D/T \rightarrow 1$**
a. Use the default parameter settings, but set the dead time to 50,000 microseconds. (This is an unphysically large value, but it will permit the simulation to run rapidly, and let you see what happens when the dead time is a large fraction of the counting time.) Do a run to completion at $t = 1000$ sec.
b. How do the **Ideal** and **GMT** bar graphs compare to the poisson curves?
c. How does the relation $1/E = 1/G - D/T$ hold up, roughly? (It's a little hard to read the graph exactly at very low counts).

6.6.3 Exercises for Series Decays

6.22 **Basic Features of Series Runs and Decay Rates**
Do a series run using all the default values. (Note that the population of three species is shown as a function of time, and the total of all three species.)
a. Make an estimate of the half-life of species 1, based on the idea of "half-life" as the time for the population to be reduced by half. Write down your method and the result. (To help reading values, you may want to zoom in on certain parts of the graph, then zoom back out again.)
b. Select **Decay Rates** from the **Pick Plot** menu. Then select **Graph It** from the top menu. This does a graph of the decay rates of each of the species, and the total decay rate. Use this plot to again roughly determine the half-life of species 1, recording the result. Your answer should be similar to that in part a.
c. What is the underlying reason for the good agreement (we hope) between parts a and b? (Your textbook will explain how the decay rate is related to the number of nuclides present. This should clarify the situation.)

6.23 **Half-Lives and Log of Decay Rates**
a. Repeat the run of Exercise 6.22. Then under **Pick Plot** select the log of decay rate and plot it. Press **L** to enter line mode, where you can move a line around the plot, and get the equation of the line. (Don't try to cross the cursors; keep the one on the left to the left of the other cursor. The program will not allow one to move past the other.) Position the line so that it lies fairly well along the curve for species 1, and press **End** (or double-click) to see the equation. Write down the equation. From the equation, determine the half-life of species 1. (To get a better view, you can omit species 3 from the plot.)
b. Does it look like this technique for finding half-lives will work for species 2? Explain why the same method will or will not work for obtaining the half-life of species 2.

6.24   **Qualitative Features of Production and Decay**

Repeat the run of Exercise 6.22 (default values).

   a. Look at the decay rate graph and see that it shows the decay rate of species 2 to be somewhat greater than species 1 after the first few data points. Is this a mistake in the program? If it's not a mistake, explain what it means.

   b. Find a case on the graph where a species has a larger population than another, but has a smaller decay rate. Describe what time this occurs, and explain the behavior. (Hint: see the half-lives of all the species.)

6.25   **Relation of Species 1 and 2 Decays**

Do a series run with default values. Study the populations and decay rates of species 1 and 2. This is not equilibrium because both species are declining with time. Could you argue that there is an "equilibrium" aspect to this behavior? How would you make a case for the "almost-equilibrium" idea?

6.26   **Effect of Numbers of Nuclei on Decay Rate**

   a. Change the initial population of species 1 to 0.2 million, and keep all the other values the same. Then do a series run. You should see the same general trend in this simulation as you did in Exercises 6.22 and 6.23. Now plot the decay rate versus time. Are there instances where the decay rate jumps up in the middle of a long decline? How can this happen if the decay rate is a purely decreasing exponential function?

   b. Repeat all of part a but change the initial population to 15 million. What effect does the change in initial population have on the decay curves?

   c. Convince yourself that larger populations behave more "smoothly" than smaller ones by running the simulation with different initial numbers. Write a concluding statement about the "truth" of the $e^{-\lambda t}$ formula for radioactive decay.

6.27   **Production and Decay of Species 2**

   a. Set the initial population of species 1 to 10 million, keeping the other defaults, and do a series run. You can see that species 2 goes through a maximum value. Try to explain how this can happen.

   b. For this same run, do a plot of decay rates versus time. Zoom in on 10 points to either side of the place where the decay rates cross from species 1 and 2 cross. Try to explain what's going on, including the slope of the curve for species 2.

6.28   **Simulated Decay From Radioactive Room Dust**

When samples of room dust are collected, lead 214 is present because it is in the decay chain from radon 222 which diffuses into the air from the walls and floor. Lead 214 decays into Bi 214 by emitting a beta-minus particle. The half-life for this decay is 26.8 minutes. Then the bismuth 214 beta decays with a half-life of 19.8 minutes. Set up this situation as a series decay (hint: leave species 3 as is) and determine what experimental half-life would be found if we started with lead 214 and sampled for a total of 3 hours.

6.29 Equilibrium Among Three Species

As a concluding "series" exercise, modify the half-lives in the series setup so that one will see "equilibrium" between species 1 and 2, and also between species 2 and 3 develop during a run which lasts 600 seconds or less. (This is the kind of thing which goes on in the alpha-decay series which trickles down to stable nuclides at or near lead.)

a. Indicate what initial population you used, and what half-lives.

b. Sketch the shapes of curves you used to convince yourself that one had a good approximation to "equilibrium" for this situation.

6.6.4 Activation Exercises

6.30 Getting Used to Activation

a. Run the default activation simulation (for silver) to completion to get an idea of what it does. Then run it again and use the hot key to turn off the flux (this corresponds to pulling the silver sample out of the howitzer) about halfway through the run. Estimate the half-life of each isotope from its population versus time. (How long does it take for the population to be cut in half?)

b. Plot the log of the decay rate. Move the left and right cursors so they surround the data after the flux was turned off, then press **End** to zoom in on this data. Go to **Line** mode (press **L** or use the hot key) and determine the equation of the line which fits i) the species 1 decay rate and ii) the species 2 decay rate. From the equations of these lines determine the decay constants (ln(2)/half-life) for species 1 and species 2. Check that the half-lives come out about right from the decay constants.

c. Now try fitting only the log of the *total* decay rate (total decays are what you would measure in an actual experiment) versus time. Try fitting a line to the data immediately after the flux was turned off, and write down the equation of this line. How good is the half-life derived from this line?

d. Repeat part c, except do the line fit for times starting about 3 minutes after the flux was turned off. How good is the half-life derived from this fit?

6.31 Varying Howitzer Time for Better Half-Life

In the previous exercise, fitting the log of the total decay rate data right after the flux was turned off did not produce an accurate result for the shorter half-life. Your job in this exercise is to modify the design of the experiment to give a better estimate of the shorter half-life right after the flux is turned off.

a. Take as an initial design goal that the decay rate from the shorter half-life must be eight times that from the longer half-life. Determine a time (flux on) in the howitzer so that when the flux is turned off there will be about eight times as many decays from the short half-life material as from the longer half-life material. Report briefly on what things you tried, what curves you looked at, and what your final results are.

b. Repeat part a, except that the design goal is to have ten times as many decays from the shorter half-life nuclide as from the longer half-life nuclide.

c. What are possible disadvantages of the shorter "flux on" times?

6.32   **Experiment Design for Activation of Copper**
Copper has two stable isotopes, just as silver does. After each has absorbed a neutron, the new nuclide is radioactive with a reasonable (in the range of seconds to hours) half-life. Design an activation experiment for copper to determine the half-lives of the two activated species from measuring the total decay rate. Because background radiation is around 0.5 counts/s, you must design so that the decay rate will be at least 1.0 counts/s throughout the experiment.

a. How long will you activate the sample before turning off the flux?

b. What flux will you use (not truly adjustable in a neutron "howitzer," but imagine that you can triple the default flux if needed)? Some data: Copper 63 neutron cross section, 4.7 barns; copper 65 cross section, 1.9 barns; copper 64 half-life, 12.75 hours; copper 66 half-life, 5.1 minutes. (There is a little more data needed; you will have to decide what this is and look it up.)

References

1. Leighton, R.B. *Principles of Modern Physics*. New York: McGraw-Hill, 1959, p. 554.

2. Blatt, F.J. *Modern Physics*. New York: McGraw-Hill, 1992, Appendix B.

3. Leighton, R.B. *Principles of Modern Physics*. New York: McGraw-Hill, 1959, p. 527.

7

Quantum Mechanics

Douglas E. Brandt

7.1 Introduction

The main topics of introductory quantum mechanics taught in most modern physics texts are Schrödinger's wave equation, its solutions, and the physical interpretation of those solutions. It is almost exclusively considered in a spatial representation, meaning that the wave function is considered a function of spatial position and time. The first problems encountered are in one spatial dimension. These are the types of problems that will be considered in this chapter, along with problems that can be reduced to one dimension. Considerable attention is usually given to solutions to the equation for potentials that allow the solution to be found in closed form relatively easily.

Schrödinger's wave equation in one dimension is a partial differential equation,

$$\frac{-\hbar^2}{2m}\frac{\partial^2 \Psi(x,t)}{\partial x^2} + V(x,t)\Psi(x,t) = i\hbar\frac{\partial \Psi(x,t)}{\partial t}. \tag{7.1}$$

The functions $\Psi(x,t)$ that satisfy this equation are called wave functions. The wave functions are related to laboratory measurements through the interpretation of the absolute square of the wave function as the probability density of particle position measurements. The wave function $\Psi(x,t)$ is related to the probability density of measuring a particle in a region of space with a particle position measurement by

$$P(a,b) = \int_a^b \Psi^*\Psi dx. \tag{7.2}$$

$P(a,b)$ is the probability that a measurement of position results in finding the particle between positions a and b, and Ψ^* is the complex conjugate of Ψ. A reasonable physical constraint on the allowed wave functions is that the probability of finding the particle somewhere in space must be identically one. This implies that the wave function be restricted to a class of functions for which

$$P(-\infty, +\infty) = \int_{-\infty}^{+\infty} \Psi^*\Psi dx = 1. \tag{7.3}$$

Equation 7.3 is called the normalization condition.

The fact that the wave function belongs to the class of functions that satisfy the normalization condition assures that the function

$$\Phi(k,t) = \frac{1}{\sqrt{2\pi}} \int_{-\infty}^{+\infty} \Psi(x,t)e^{-ikx}\,dx \tag{7.4}$$

exists. This function is called the Fourier transform of $\Psi(x,t)$. It has a relatively simple interpretation. Any wave function can be constructed from a linear combination of harmonic waves. The function $\Phi(k,t)$ specifies the linear combination of harmonic waves that is needed to construct the wave function of interest. The variable k is the wave number related to the wavelength λ by

$$k = \frac{2\pi}{\lambda}. \tag{7.5}$$

and the direction of the wave velocity determines the sign of k. Specifying $\Phi(k,t)$ is equivalent to specifying $\Psi(x,t)$ because the Fourier transform is unique and includes all the information necessary to construct the wave function. It is relatively straightforward to write the equation that $\Phi(k,t)$ satisfies that is the equivalent of Schrödinger's equation.[1]

Because the dynamical variable of interest is usually the momentum and the deBroglie relation states that

$$p = \hbar k, \tag{7.6}$$

the function $\Phi(k,t)$ is usually written with a change of variables to $\Phi(p,t)$. The function $\Phi(p,t)$ is called the momentum-space wave function. In terms of the probabilistic interpretation of the wave function, the probability of measuring a momentum between p_1 and p_2 is given by

$$P(p_1,p_2) = \int_{p_1}^{p_2} \Phi^*\Phi\,dp. \tag{7.7}$$

7.1.1   Uncertainty Principle

A direct consequence of the Fourier transform relationship between the spatial and momentum space wave functions is the Heisenberg uncertainty principle. The second moment about the mean, σ^2, of a function is defined by

$$\sigma^2 = \int_{-\infty}^{\infty} (x - \bar{x})f(x)\,dx, \tag{7.8}$$

where $\bar{x}$ is the first moment or mean of the function

$$\bar{x} = \int_{-\infty}^{\infty} xf(x)\,dx. \tag{7.9}$$

It is an elementary result of complex analysis that the product of second moments about the mean (variance) of a function and its Fourier transform has a lower limit.[2] If the square root of the second moment about the mean of the probability density, σ_x, is adopted as the measure of uncertainty in the location of a particle and σ_p is adopted as the measure of uncertainty in momentum of the particle, then the general result from complex analysis becomes

$$\sigma_x\sigma_p \geq \frac{\hbar}{2}. \tag{7.10}$$

The equality holds for gaussian distributions of probability density.

7.1.2   Time Evolution of Wave Functions

Schrödinger's equation in one dimension (Eq. 7.1) is a partial differential equation that determines the time evolution of the wave function. Partial differential equations in general can be very difficult to solve. If the potential energy function, $V(x,t)$, only depends on space and not time, $V(x,t) = V(x)$, then Schrödinger's equation is separable into two ordinary differential equations:

$$i\hbar \frac{dT(t)}{dt} = ET(t) \tag{7.11}$$

$$\frac{-\hbar^2}{2m} \frac{d^2\psi(x)}{dx^2} + V(x)\psi(x) = E\psi(x), \tag{7.12}$$

where

$$\Psi(x,t) = \psi(x)T(t), \tag{7.13}$$

and E is called the separation constant and corresponds to the energy of the wave function. The solution of Eq. 7.11 is elementary. The solutions are given by

$$T(t) = Ce^{-iEt/\hbar}, \tag{7.14}$$

where C is some constant. Equation 7.12 is called the time-independent Schrödinger equation. The solution to the time-independent Schrödinger equation requires more work than Eq. 7.11 in all but the free particle problem, $V = 0$ everywhere.

7.1.3   Free Particle Solutions

In the free particle case, the time-independent Schrödinger equation reduces to

$$-\frac{\hbar^2}{2m} \frac{d^2\psi(x)}{dx^2} = E\psi(x), \tag{7.15}$$

which has solutions of the form

$$\psi(x) = Ae^{\kappa x} + Be^{-\kappa x}, \tag{7.16}$$

where

$$\kappa = \frac{\sqrt{-2mE}}{\hbar}. \tag{7.17}$$

The solutions of physical interest are those with positive values of E. It should be obvious that the solutions with a negative value of E will have a divergent behavior at either large negative values of x or large positive values of x. This is consistent with the kinetic energy being a non-negative quantity. If E is positive, the solutions for $\Psi(x,t)$ can be rewritten in the form

$$\Psi(x,t) = \psi(x)T(t) = (A'e^{ikx} + B'e^{-ikx})e^{-iwt}$$
$$= A'e^{i(kx-\omega t)} + B'e^{-i(kx+\omega t)}, \tag{7.18}$$

where

$$k = \frac{\sqrt{2mE}}{\hbar} \tag{7.19}$$

and

$$\omega = \frac{E}{\hbar}. \tag{7.20}$$

These solutions are the sum of terms that describe a wave of amplitude A' traveling in the $+x$ direction and a wave of amplitude B' traveling in the $-x$ direction.

There is a problem with the interpretation of the wave function as the probability density for these solutions, but it is a physically reasonable problem to have. A calculation of the probability density shows

$$\Psi^*\Psi = A'^*A' + B'^*B' + A'B'^*e^{2ikx} + A'^*B'e^{-2ikx}. \tag{7.21}$$

When an attempt is made to determine A' and B' so that Ψ satisfies the normalization condition, it is found that these solutions cannot satisfy the normalization condition as the integral in Eq. 7.3 diverges. It is not so surprising that a problem would be encountered in this case. Consider the physical problem represented by this situation. The particle is free to be anywhere with equal probability, so the probability of finding the particle in any finite interval of space must be zero.

At this point, consider the time evolution of these solutions. The only thing that changes with time is the multiplicative factor $T(t)$. This factor only changes the overall phase of the wave function. It doesn't change the amplitude or relative phase of the wave function in different parts of space. This implies that the probability density remains constant in time,

$$\Psi^*\Psi = \psi^*(x)T^*(t)\psi(x)T(t) = \psi^*(x)\psi(x)e^{\frac{+iEt}{\hbar}}e^{\frac{-iEt}{\hbar}} = \psi^*(x)\psi(x). \tag{7.22}$$

These solutions are not very interesting in themselves, as there is no time evolution of any measurable quantity.

Both the above problems of these solutions can be fixed once it is realized that Schrödinger's equation is a *linear* differential equation. One of the first properties of linear differential equations that is taught is that sums of solutions of a linear differential equation are also solutions to the same differential equation. The solutions of interest might be wave functions that correspond to a high probability density near a particular point in space at a particular time. Linear combinations of the plane wave solutions can be found that have this behavior. These linear combinations can satisfy the normalization condition (Eq. 7.3). Because these linear combinations will not be factorable into a function of position and a function of time, there will also be some nontrivial time dependence of the probability density.

Wavepackets can be written as a linear combination of plane waves. Wave packets that have a probability close to one of being in some specific finite region of space are said to be *localized*. The paragraph above implies that wave packets will have some nontrivial time dependence. A useful way to understand the motion of wave packets is through the phase velocity, v_p, and the group velocity, v_g. The phase velocity is defined as

$$v_p = \frac{\omega}{k}. \tag{7.23}$$

Why is this called the phase velocity? Look at the general form of a harmonic traveling wave,

$$\Psi = Ae^{i(kx-\omega t)} \tag{7.24}$$

The argument of the exponential, $kx - \omega t$, can be considered a phase angle. If x is changed with time to keep this phase angle constant, it must change at a rate of v_p. This is the velocity that a given plane wave component advances in time.

If the phase velocity is the same for all plane wave components of a wavepacket, the wavepacket moves forward with the phase velocity and does not change shape with time. If the phase velocity is not the same for all the plane wave components, one might guess that the wave packet moves forward with the mean velocity of the components. This is not what happens. The wavepacket moves forward with the group velocity, v_g, defined as

$$v_g = \frac{d\omega}{dk}. \tag{7.25}$$

To see why this may be true,[3] consider a wave that is the sum of two component waves of equal amplitude and nearly the same wavelength,

$$\Psi = Ae^{i(k_1x-\omega_1 t)} + Ae^{i(k_2x-\omega_2 t)}. \tag{7.26}$$

Graphing the real part of the component waves and their linear combination in Figure 7.1, the resulting wave appears to be an amplitude modulated harmonic wave. (Note that this particular linear combination is not normalizable, but is sufficient to demonstrate the concept of group velocity. The type of linear combination of solutions that is normalizable must be written as a weighted integral of plane wave solutions.) This result is can be expected if a factor

$$\exp[i(\frac{k_1 + k_2}{2} - \frac{\omega_1 + \omega_2}{2})] \tag{7.27}$$

is factored out of each term in Eq. 7.26. The result is

$$\Psi = A \exp[i(\frac{k_1 + k_2}{2} - \frac{\omega_1 + \omega_2}{2})]$$
$$(\exp[i(-\frac{\Delta k}{2}x + \frac{\Delta \omega}{2}t)] + \exp[i(\frac{\Delta k}{2}x - \frac{\Delta \omega}{2}t)]), \tag{7.28}$$

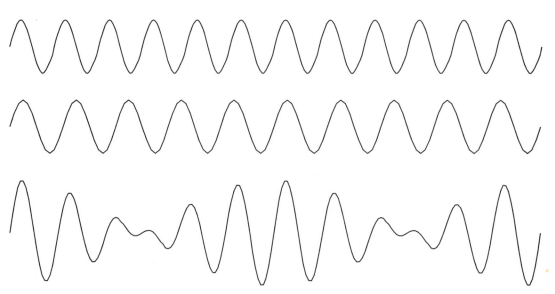

Figure 7.1: The two upper graphs display the real part of two plane waves. The second plane wave has a wavelength that is 6/5 of the wavelength of the first plane wave. The lowest graph shows the linear combination of those two waves.

where $\Delta k = k_2 - k_1$ and $\Delta \omega = \omega_2 - \omega_1$. Equation 7.28 can be rewritten

$$\Psi = 2A \exp[i(\frac{k_1 + k_2}{2} - \frac{\omega_1 + \omega_2}{2})]\cos(\frac{\Delta k}{2}x - \frac{\Delta \omega}{2}t). \qquad (7.29)$$

The exponential factor describes a harmonic wave that moves with a phase velocity $\bar{\omega}/\bar{k}$, where $\bar{\omega}$ and $\bar{k}$ are the mean of ω_1 and ω_2 and of k_1 and k_2, respectively. The cosine factor describes the amplitude modulation seen in Figure 7.1. The probability density of this wave is given by

$$\Psi^*\Psi = 4|A|^2 \cos^2(\frac{\Delta k}{2}x - \frac{\Delta \omega}{2}t). \qquad (7.30)$$

The velocity of the phase of this cosine term is

$$v = \frac{\Delta \omega/2}{\Delta k/2} = \frac{\Delta \omega}{\Delta k}. \qquad (7.31)$$

The "lumps" of probability density move with this velocity. In the limit of a continuous distribution of k values included in the wavepacket, $\Delta \omega/\Delta k$ becomes $d\omega/dk$.

Therefore, the motion of wavepackets can be understood from the relationship of ω and k in many situations. The dependence of ω on k is called the dispersion relation. The dispersion relation depends on the wave equation. The most familiar time evolution of wavepackets is that of electromagnetic waves in vacuum and acoustic waves at low frequencies. The dispersion relation for these waves is given by

$$\omega = ck, \qquad (7.32)$$

where c is the wave speed and is the same for all frequencies. It directly follow that the group velocity is also c. Because c is the same for all plane wave components of a wavepacket, the wavepacket evolves by moving with the speed c without changing shape.

Using Eqs. 7.19 and 7.20, it is easy to solve for the dispersion relation for the wavefunction that represents free particles evolving according to Schrödinger's equation,

$$\omega = \frac{\hbar}{2m}k^2, \tag{7.33}$$

resulting in a phase velocity

$$v_p = \frac{\omega}{k} = \frac{\hbar k}{2m}. \tag{7.34}$$

The group velocity derived from this dispersion relation is

$$v_g = \frac{d\omega}{dk} = \frac{\hbar k}{m}. \tag{7.35}$$

The group velocity is twice the phase velocity!

7.1.4 Piecewise Constant Potentials

The next easiest class of time-independent Schrödinger equations to solve are those that have potentials that are piecewise constant. This means that space can be broken into a countable number of intervals on which the potential is a constant over the entire interval. First, a single well or barrier potential will be considered and then the method will be expanded to include potentials of an arbitrary number of piecewise constant potential regions.

The single well or barrier made of a piecewise constant potential has space divided into three regions as shown in Figure 7.2. Region 1 extends over $-\infty < x < 0$, and the potential in that region is V_1. Region 2 extends over $0 < x < L$,

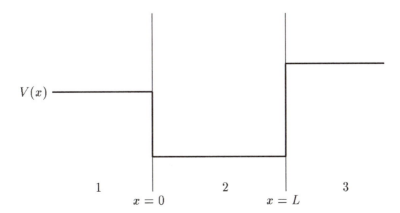

Figure 7.2: Numbering of regions in a finite square well problem.

and the potential in that region is V_2. Region 3 extends over $L < x < \infty$, and the potential in that region is V_3. The solution in each region is given by

$$\psi_1(x) = A_1 e^{\kappa_1 x} + B_1 e^{-\kappa_1 x} \tag{7.36}$$

$$\psi_2(x) = A_2 e^{\kappa_2 x} + B_2 e^{-\kappa_2 x} \tag{7.37}$$

$$\psi_3(x) = A_3 e^{\kappa_3 x} + B_3 e^{-\kappa_3 x}, \tag{7.38}$$

where

$$\kappa_n = \frac{\sqrt{2m(V_n - E)}}{\hbar}, \tag{7.39}$$

and E is the value of the separation constant (energy) for a particular solution.

There are seven unknowns in the general solution stated above: three A's, three B's and E. Equation 7.3, the normalizability condition, and Schrödinger's equation together imply that the wavefunction is continuous everywhere and the derivative of the wavefunction is continuous except at infinite discontinuities in the potential function.[1] In this case, continuity of the wavefunction implies

$$\psi_1(0) = \psi_2(0) \tag{7.40}$$

$$\psi_2(L) = \psi_3(L), \tag{7.41}$$

and continuity of the derivative of the wavefunction implies that

$$\left.\frac{d\psi_1}{dx}\right|_0 = \left.\frac{d\psi_2}{dx}\right|_0 \tag{7.42}$$

$$\left.\frac{d\psi_2}{dx}\right|_L = \left.\frac{d\psi_3}{dx}\right|_L. \tag{7.43}$$

Rewritten in terms of the explicit solutions, these boundary conditions become

$$A_1 + B_1 = A_2 + B_2 \tag{7.44}$$

$$A_2 e^{\kappa_2 L} + B_2 e^{-\kappa_2 L} = A_3 e^{\kappa_3 L} + B_3 e^{-\kappa_3 L} \tag{7.45}$$

$$\kappa_1 A_1 - \kappa_1 B_1 = \kappa_2 A_2 - \kappa_2 B_2 \tag{7.46}$$

$$\kappa_2 A_2 e^{\kappa_2 L} - \kappa_2 B_2 e^{\kappa_2 L} = \kappa_3 A_3 e^{\kappa_3 L} - \kappa - 3B_3 e^{-\kappa_3 L}. \tag{7.47}$$

If the problem is a potential well problem, there will be both bound state and scattering state solutions. Bound state solutions will have values of the energy, E, less than the minimum of V_1 and V_3 and greater than V_2, making κ_1 and κ_3 real numbers and κ_2 an imaginary number. To avoid some of the confusion caused by the imaginary parameter κ_2, the variable ik_2 will be substituted for κ_2 so that real and imaginary parts will appear explicitly real and imaginary, respectively. If the bound state solutions are desired, the normalizability condition implies that B_1 and A_3 are equal to zero, otherwise the solution will be an increasing exponential function as x becomes a large negative or large positive number. This reduces the number of unknowns to five. A fifth equation, in addition to the boundary conditions stated in Eq. 7.40, is the normalization condition. The four boundary conditions in Eq. 7.40 can be reduced by elimination to the single equation

$$(ik_2 + \kappa_1)(ik_2 + \kappa_3)e^{ik_2L} = (ik_2 - \kappa_1)(ik_2 - \kappa_3)e^{-ik_2L}. \tag{7.48}$$

Expanding this equation and writing the exponentials in terms of cosine and sine functions results in

$$k_2(k_1 + k_3)\cos k_2L + (k_1k_3 - k_2^2)\sin k_2L$$
$$= -k_2(k_1 + k_3)\cos k_2L - (k_1k_3 - k_2^2)\sin k_2L. \tag{7.49}$$

Rearranging Eq. 7.49,

$$\cot k_2L - \frac{k_1k_3 - k_2^2}{k_2(k_1 + k_3)} = 0. \tag{7.50}$$

Remembering that the values of κ_1, κ_3, and k_2 are functions of E, it is seen that the Eq. 7.50 is an equation containing the single unknown E. The solution of this equation gives the allowed values of E for which physically reasonable solutions (those that are well behaved for large negative and positive values of x) exist for the Schrödinger equation. Note that this equation is a transcendental equation. The solution of this equation is therefore a good problem to attempt to solve numerically with computer.

Once the values of E that satisfy Eq. 7.50 are found, the constants A_1, A_2, B_2, B_3 can be found from the boundary condition equations and the normalization condition. Writing the normalization condition in terms of the explicit solution for the wavefunction,

$$\int_{-\infty}^{+\infty} \psi^*\psi\,dx = \int_{-\infty}^{0} |A_1|^2 e^{\kappa_1 x}\,dx + \int_{0}^{L} |A_2|^2 + |B_2|^2$$
$$+ A_2^*B_2 e^{-2ik_2x} + A_2B_2^* e^{+2ik_2x}\,dx + \int_{L}^{+\infty} e^{-\kappa_3 x}\,dx. \tag{7.51}$$

The integration can easily be completed to simplify this condition to

$$\frac{|A_1|^2}{2\kappa_1} + |A_2|^2 L + |B_2|^2 L + \frac{A_2^*B_2}{-2i}(e^{-2ik_2L} - 1)$$
$$+ \frac{A_2B_2^*}{2i}(e^{+2ik_2L} - 1) + \frac{|B_3|^2}{2\kappa_3} = 1. \tag{7.52}$$

Equation 7.52 along with any three of the boundary condition equations is a set of four independent equations for the four variables A_1, A_2, B_2, and B_3. The normalization condition is not a linear equation in the unknown variables, so the solution for these constants is not unique. Any solution for A_1, A_2, B_2, and B_3 can be used to generate other solutions by multiplying each of these constants by a factor of $\exp i\theta$ where θ is a real number. Often, this is stated that the solution is unique up to an arbitrary phase factor. Usually the solution for which ψ is real is the one that is quoted.

The solutions will be scattering solutions if the value of the energy E is greater than either V_1 or V_3. In fact, there will be solutions to the time-independent Schrödinger equation for all values of E greater than the minimum of V_1 and V_3. In this case, the quantities of physical interest are the reflection and transmission coefficients of the barrier. In two or three dimensions, the reflection and transmission coefficients generalize to the scattering differential cross section.[4] Without any loss of generality, assume that E is greater than the potential energy in region 1. This implies that κ_1 is imaginary and as in the bound state discussion, κ_1 will be written as ik_1 explicitly to show the imaginary quantity. The component of the solution $A_1 \exp(ik_1 x)$ in region 1 is a traveling wave with a velocity in the positive x direction. The component of the solution $B_1 \exp(-ikx)$ is traveling wave with a velocity in the $-x$ direction.

The potentials in region 3 could be greater than or less than the value of E for the desired solution. If V_3 is greater than E, the same argument used in the bound state solution discussion implies that A_3 is zero. If V_3 is less than E, the term $B_3 \exp(-\kappa x)$ of the solution in region 3 will be a plane wave traveling in $-x$ direction. The physical application of this situation is to an experiment where particles are sent toward a potential barrier from region one and observations are made after the particle has interacted with the barrier. There should be no component of the solution that is a wave traveling toward the barrier in region three. This implies that B_3 is zero when E is greater than V_3.

The probability current (see exercise 7.21) reflection coefficient will be the ratio $|B_1|^2/|A_1|^2$. The transmission coefficient will be given by $(|A_3|^2 k_3)/(|A_1|^2 k_3)$. The problem is often simplified by setting A_1 equal to 1. This leaves the four unknowns B_1, A_2, B_2, and either A_3 or B_3 for which to solve. There are four boundary conditions giving a system of four unknowns and four equations linear in these unknowns. This system is easy solved by elimination to determine the reflection and transmission coefficients.[1]

The above discussions are easily generalized to the case of a potential made up of n piecewise constant regions. To keep the picture simple and ordered in a familiar way, number the regions consecutively with integers n, starting at 1 for the leftmost region and increasing n to the right as in Figure 7.3. In the nth region, the Schrödinger equation takes the form

$$\frac{\hbar^2}{2m} \frac{d^2\psi(x)}{dx^2} + V_n \psi(x) = E\psi(x), \tag{7.53}$$

where V_n is the potential in the nth region. The solution in each region takes the form

$$\psi_n(x) = A_n e^{kx} + B_n e^{-kx}, \tag{7.54}$$

where

$$k = \frac{\sqrt{2m(V_n - E)}}{\hbar}. \tag{7.55}$$

Again, with the interpretation of $|\Psi|^2$ as the probability density, Schrödinger's equation requires that the wavefunction be continuous everywhere and that its

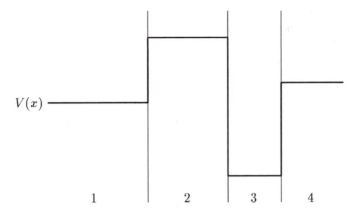

$V(x)$

1 2 3 4

Figure 7.3: Numbering of regions in a piecewise constant potential problem.

derivative be continuous everywhere except at infinite discontinuities in the potential. That implies the following conditions at each boundary between two regions of constant potential:

$$\psi_n(x_b) = \psi_{n-1}(x_b) \tag{7.56}$$

$$\frac{d\psi_n}{dx}\bigg|_{x_b} = \frac{d\psi_{n-1}}{dx}\bigg|_{x_b}, \tag{7.57}$$

where x_b is the boundary between region $n-1$ and region n and ψ_{n-1} and ψ_n are the solutions in the corresponding regions. If there are total of n regions, then there are $n-1$ boundaries. This results in $2(n-1)$ equations. There are $2n$ unknown A_n and B_n coefficients of the solutions and the value of the energy, E, is unknown at this point. How the problem is treated depends on whether the energy of the solution is less than or greater than the potential in the first and last region. If the energy of interest is less than the potential in the first and last region, the solution is a bound state solution. If the energy of interest is greater than either of the potentials in the first or last region, the problem is solved as a scattering problem.

The problem of solving Schrödinger's equation for bound states results in determining the energies for which the wavefunction is normalizable. There are solutions to Schrödinger's equation for all values of E, but only those that are square integrable (i.e., the integral from minus infinity to infinity of the absolute square of ψ is finite) are consistent with the interpretation of the absolute square of ψ as a probability density. This is true for any bound state problem and not just the piecewise constant potential problem. In the piecewise constant potential problem, the solutions in the first and last regions are seen to be exponential functions of x with real coefficients in the exponential because V is greater than E in these regions. Because the rightmost region includes arbitrarily large negative x in it, the exponential term with the negative coefficient must not appear or the wavefunction will not be square-integrable. In other words, $B_1 = 0$. Similarly, in the last region the term with exponential with a positive coefficient must not appear, $A_n = 0$. This now reduces the problem to $2n - 2$ unknown A's and B's and an unknown

E. The $2n - 2$ boundary condition equations and the normalization condition give as many equations as unknowns and the system of equations can be solved.

When the energy is greater than the potential in either or both of the first or last regions, the solution to Schrödinger's equation is in the form of a traveling wave in those regions. There are then solutions to Schrödinger's equation for all values of the energy E, and none of these solutions is normalizable. However, in this case the meaning of the normalizability of the wavefunction must be looked at more closely. The laboratory experiment that this problem would model is one in which particles are sent incident on a scattering potential and then detected after interacting with the scattering potential. Although the wavefunction is not normalizable, there is useful information in the solution to Schrödinger's equation from the relative amplitudes of the traveling wave solutions that are incident and retreating from the scattering potential. These relative amplitudes remain finite and satisfy conservation of particle flux despite the fact that the wavefunction cannot be normalized.

Relating this to the mathematical problem for the piecewise constant potential, the energy E and the incident amplitude A_1 are chosen. If the potential in the last region is greater than the energy, then A_n must be zero or the result corresponds to arbitrarily large relative probability of finding the particle in a region not classically allowed. If the energy is greater than the potential in the final region, then B_n must be zero to correspond to the laboratory model of no particles incident on the scattering potential from the right. In either case, specify three of the unknown constants in the problem leaves the $2n - 2$ A's and B's yet undetermined, and the $2n - 2$ boundary conditions relating them. The unknown A's and B's can be determined. Relating this to the physics of the laboratory problem, the amplitude of the solutions that are outgoing waves from the region of the barrier are related to the number of particles reflected from and transmitted through the barrier.

The reflection coefficient is the ratio of the probability flux of the reflected wave to the ratio of the probability flux of the incident wave. The transmission coefficient is the ratio of the probability flux in the region beyond the barrier to the probability flux incident on the barrier. The probability flux would classically be defined as the probability density multiplied by its velocity. In quantum mechanics, the probability flux can be shown to be

$$j = \frac{i\hbar}{2m}(\psi\frac{d}{dx}\psi^* - \psi^*\frac{d}{dx}\psi). \tag{7.58}$$

This result is a direct consequence of Schrödinger's equation and conservation of probability (see Exercise 7.21). From the given solutions for the wavefunction in regions of piecewise constant potential, it is easy to show (see Exercise 7.22) that in a region where the wavefunction is given by

$$\psi = Ae^{i(kx-\omega t)} + Be^{-i(kx-\omega t)}, \tag{7.59}$$

the probability flux is given by

$$j = (A^*A - B^*B)\frac{\hbar k}{m} \tag{7.60}$$

when k is real.

7.2 *Numerical Solutions for Bound States*

There are many numerical methods for finding solutions to Schrödinger's equation for arbitrary potentials. The most direct is to write the discrete version of Schrödinger's time-independent equation and numerically integrate to determine the wavefunction for a given energy value. The energy value is searched for wavefunctions that are square integrable (normalizable).

As Schrödinger's time-independent equation contains a term proportional to the second derivative of the wavefunction, a discrete approximation to the second derivative is needed. Given a function $f(x)$, an approximation to the derivative of $f(x)$ is

$$\frac{df(x)}{dx} \approx \frac{f(x) - f(x - \Delta x)}{\Delta x} \tag{7.61}$$

when Δx is "small." The second derivative of $f(x)$ is the derivative of $\frac{df(x)}{dx}$ so an approximation to the second derivative of $f(x)$ is

$$\frac{d^2 f(x)}{dx^2} \approx \frac{f'(x + \Delta x) - f'(x)}{\Delta x}. \tag{7.62}$$

One way to proceed at this point is to cover the region of interest in the problem with a grid of equally spaced points. This grid is then numbered sequentially from the smallest x-value to the highest with integers. The above derivative approximations can then be written

$$\frac{df(x)}{dx} = \frac{f(x_n) - f(x_{n-1})}{\Delta x} \tag{7.63}$$

and

$$\frac{d^2 f(x)}{dx^2} = \frac{[f(x_n) - f(x_{n-1})]/\Delta x - [f(x_{n-1}) - f(x_{n-2})]/\Delta x}{\Delta x}$$
$$= \frac{f(x_n) + f(x_{n-2}) - 2f(x_{n-1})}{(\Delta x)^2}. \tag{7.64}$$

Schrödinger's time-independent equation written in terms of this approximate second derivative becomes

$$\frac{-\hbar^2}{2m} \frac{\psi(x_n) + \psi(x_{n-2}) - 2\psi(x_{n-1})}{(\Delta x)^2} + V(x_{n-1})\psi(x_{n-1}) = E\psi(x_{n-1}). \tag{7.65}$$

This equation can be solved for $\psi(x_n)$:

$$\psi(x_n) = \frac{2m}{hbar^2}(V(x_{n-1}) - E)(\Delta x)^2 - \psi(x_{n-2}) + 2\psi(x_{n-1}). \tag{7.66}$$

This is a recursion relation that can be used to generate values for $\psi(x_n)$ given values of $\psi(x_1)$ and $\psi(x_2)$. Just as in the case of piecewise constant potentials, to ensure the normalizability of the wavefunction the value of $\psi(x)$ must go to zero at arbitrarily large values of x. This implies that $\psi(x_n)$ must approach zero as n becomes large. This condition is only satisfied for particular values of the separation constant, E. To determine the bound state energy eigenvalues and the wave functions associated with those eigenvalues, a search is made for those values of E that lead to normalizability of the wavefunction.

7.3   The Program

The program QUANTUM consists of four different sections.

7.3.1   Uncertainty Principle

The first section, **Uncertainty Principle**, is designed to illustrate the reciprocal
nature of the width of the spatial wavefunction and the width of the distribution of
momentum of the wavefunction. Of course, this is just the Heisenberg uncertainty
principle. The more mathematically sophisticated users will recognize this as
a general property of the nature of Fourier transform pairs of functions. It is
independent of the fact that the waves that are being studied are the solution
to the Schrödinger equation or any other wave equation.

The program allows for the adjustment of mean position and momentum, the
standard deviation of the position or momentum, and the type of wave function
for one of the distributions. As shown in Figure 7.4, the mean and standard
deviation parameters are adjustable by sliders located below the graphs of the
two distributions. To change the type of a distribution, select **Wave Function**
parameters under the **Parameters** menu.

7.3.2   Dispersion

The second section **Dispersion** is designed for studying the time development of
wave packets and to illustrate the ideas of group and phase velocity. A wave packet
is developed in time according to the dispersion relation selectable from several
types or the user may input a dispersion relation with which the wave packet's
development is determined.

The display for this section of the program is shown in Figure 7.5. There are
two sliders that allow the adjustment of the mean wave vector and the initial width
in position of the wavepacket. The momentum distribution of the wavepacket is
shown in the center graph on the right side of the display. The dispersion relation
for the wavepacket is displayed in the graph at the lower right of the display. The
wavefunction is shown in the graph on the right side of the display. The dispersion
relation used to evolve the wavepacket can be chosen by selecting **Dispersion**
under the **Parameters** menu.

7.3.3   Barriers

The third section, **Barriers**, is for studying the solutions to the Schrödinger equation
when the solution is not bound. A piecewise constant barrier is adjustable by the
user. The program uses exact solutions to the piecewise constant wavefunction
and numerically solves the boundary condition equations for the amplitudes in the
different regions. The display for this section of the program is shown in Figure 7.6.
The program shows the reflection and transmission coefficients of the barrier for a
range of energies in the graph at the upper right of the display. In this same graph,
the energy distribution of the current wavepacket is displayed. There is a slider at
the lower right side of the display to allow the adjustment of the energy of mean

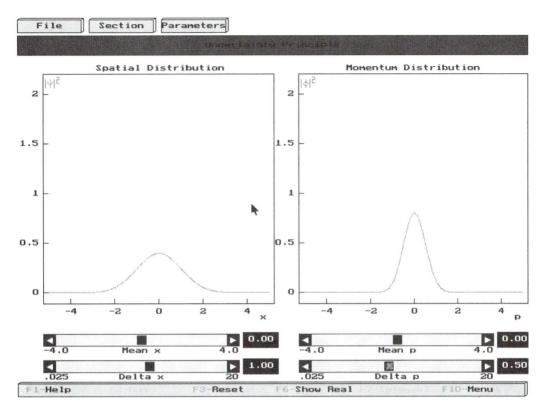

Figure 7.4: This is the display for the uncertainty principle section of program. The left graph shows the spatial wavefunction and the right graph shows the momentum wavefunction. The sliders below the graph allow for the adjustment of the mean position, mean momentum, uncertainty of position, and uncertainty of momentum.

wavevector of the wavepacket. The graph on the left side of the display shows the time-dependent behavior of both energy eigenfunctions and wavepackets incident upon the barrier.

7.3.4 Bound Particles

The fourth section, **Bound Particles**, is for studying solutions to the Schrödinger equation that are bound, states i.e., confined to a particular region of space. The display shows the energy eigenvalue solutions and the time development of a wavepacket that consist of a linear combination of bound energy eigenstates. The particular linear combination shown is selectable by the user. The potential is selectable by the user.

The display for the bound solution section of the program is shown in Figure 7.7.

One possible selection for the potential is a piecewise constant potential with widths of regions and potential selectable by the user. In this case, the program uses the exact solutions for piecewise constant potentials for the wave function and determines the energy eigenvalues from a numerical solution of the exact

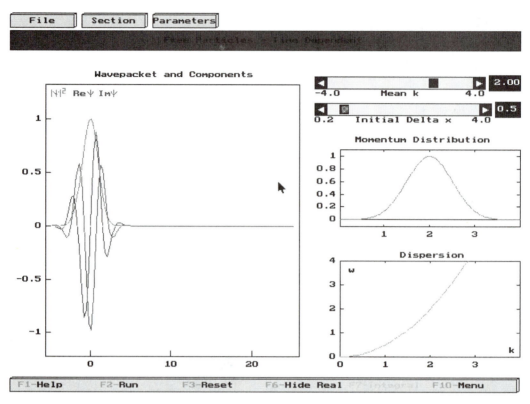

Figure 7.5: This is the display for dispersion section of program. The graph on the left half of the display shows the wavefunction and probability density. The sliders at the top left of the display allow for the adjustment of the mean *k* and initial uncertainty in position of the wave packet. The top graph on the right side shows the momentum distribution of the wave packet, and the bottom graph on the right shows the dispersion relation by which the wave packet evolves.

boundary value equations. The user can also input arbitrary potentials that may have bound solutions. In this case the program does a search for energy eigenvalues by integrating the time independent Schrödinger equation and looking for the proper wave function behavior for positions far from the potential well.

7.4   Details of the Program

7.4.1   Program Control

The program is controlled by making selections from the menus and function keys and by entering appropriate information requested by the program.

There are four main menus each with several choices:

- **File**:

    — **About CUPS**: This menu item displays information about the CUPS project.

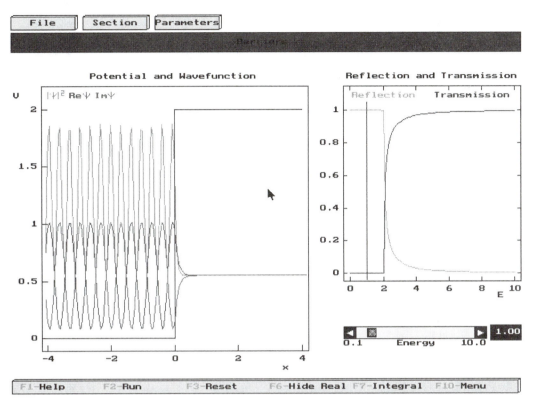

Figure 7.6: This is the display for the barrier section of program. The graph on the left side of the display shows the wavefunction, probability density, and potential. The graph on the right side of the display shows the reflection and transmission coefficients as a function of energy for the barrier. The slider allows the adjustment of the energy of the eigenfunction or the mean energy of the wave packet.

— **About Program**: This menu item displays information about the program.
— **About Section**: This item displays information about the current section of the program.
— **Configuration**: This menu item allows for user selection of temporary file path directory, colors, double-click time of the mouse, and delay time. It also allows the user to view the amount memory available on the heap.
— **Quit**: This menu item exits the program and returns to the operating system.

• **Section**:

— **Uncertainty Principle**: This menu item starts the uncertainty principle investigation section of the program.
— **Time-Dependent Free Particles**: This menu item starts the time-dependent free particle investigation section of the program.
— **Barriers**: This menu item starts the barrier investigation section of the program.

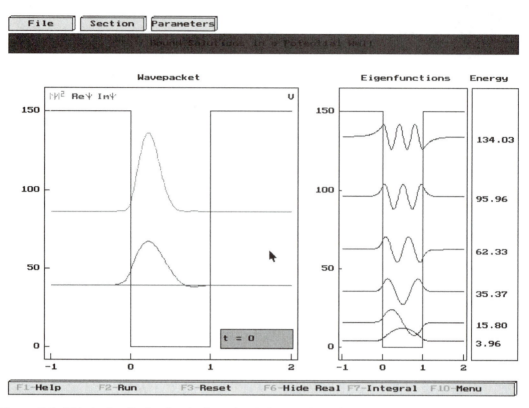

Figure 7.7: This is the display for the bound solution section of program. The left graph displays the wavefunction, the probability density, and the potential. The right graph displays the potential and the calculated energy eigenfunctions. The right column of numbers are the energy eigenvalues of the corresponding eigenfunction on the right graph.

— **Bound Particles in a Well**: This menu item starts the bound particles in a well investigation section of the program.

- **Parameters**

  — **Wave Function Parameters**: This menu item allows one to choose parameters of the problem under investigation that relate to the wavepacket displayed.

  — **Barrier Parameters**: This menu item allows one to construct a barrier of up to five different regions of constant potential. It is only active in the **Barriers** section of the program.

  — **Well Parameters**: This menu item allows one to choose a potential well. A square well for which the program determines exact solutions or a user defined potential can be selected. This item is only active in the **Bound Solutions in a Well** section of the program.

- **Units**:

  — **nm, eV, fs**: This menu item sets the units used in the program to nanometers for length, electron volts for energy, and femtoseconds for time. This implies that the value of $\hbar$ is 0.6582 eV-fs.

— **Dimensionless, $\hbar = 1$, electron mass $= 1$**: This menu item selects the units used in the program to be a dimensionless set of units with $\hbar = 1$ and electron mass $= 1$.

— **About Units**: The selection of this menu item brings up a screen that displays information about the selection of a set of units.

• Hot Keys: There are six hot keys for controlling the program:

• F1-Help: This hot key displays a help screen relevant to the current section of the program being used.

• F2-Run/Stop: This hot key starts and stop the animation in time of the wavepackets displayed on the screen. It is active in all sections except the UNCERTAINTY PRINCIPLE section of the program.

• F3-Reset: This hot key resets the time to zero in all of the sections that show time evolution of wavepackets. It is not active in the UNCERTAINTY PRINCIPLE section of the program.

• F6-Show Real: This hot key toggles the display of the real and imaginary components of the wavepackets on and off.

• F7-Integral: This hot key starts the integration tool of the program. It is only active in the **Barriers** and **BOUND SOLUTIONS IN A WELL** sections of the program.

• F10-Menu: This hot key will activate the menu.

7.5 Structure of the Program

The program consists of the following units:

• QUANTUM.PAS: This is the main program and its main functions are to take care of initializations, to switch back and forth between the various sections, and to clean up when the program is exited.

• QM_FPTI.PAS: This file contains the procedure for the **Uncertainty Principle** section of the program.

• QM_FPTD.PAS: This file contains the procedure for the **Free Particle Time-Dependent** section of the program.

• QM_BARR.PAS: This file contains the procedure for the **Barriers** section of the program.

• QM_SWEL.PAS: This file contains the procedure for the **Bound Solutions in a Well** section of the program.

- QM_IFAC.PAS: This file contains the procedures for handling the events generated by the user such as mouse clicks and keyboard input. It sends messages to the section procedures containing the information about the events that occurred.

- QM_FUNC.PAS: This file contains functions used by the various sections of the programs that needed to be accessed externally.

- OBJECTS.PAS: This file defines some of the objects that are used in this program such as a complex extension of the **DVectors** defined in CUPSPROC.

- XSLIDERS.PAS: This file contains an extension to the CUPS sliders object that allows logarithmic sliders to be defined.

The program also requires the CUPS units CUPS, CUPSMUPP, CUPSPROC, CUPSFUNC, CUPSGUI, and CUPSGRPH. The program requires the Turbo Pascal units **Crt**, **Dos**, and **Graph**.

7.6   Units

A set of dimensionless units can be selected by the user. The physical quantities involved in this program are energy, momentum, distance, mass, and time. There are three independent units in a consistent set of units that can be chosen to measure these five quantities. In the programs, $\hbar$ and the electron mass are always identically equal to one in this dimensionless set of units. This leaves one more independent unit that can be chosen by the user of the program. After that unit is selected, the relationship of the units of the quantities found in the program to established systems of units can then be determined.

As an example, suppose the unit of length is chosen to be a nanometer. The units of momentum are the units of $\hbar$ divided by the units of length. This implies the units of momentum will be $1/nm$. The units of energy are units of momentum squared divided by the units of mass. That implies that for the given dimensionless quantities and length units of nanometers that the units of energy will be $1/nm^2$. The units of time are the units of $\hbar$ divided by the units of energy, so the units of time must be nm^2.

7.7   Exercises

7.1   Uncertainty Principle

Run the program QUANTUM. It starts up in the uncertainty principle section of the program and displays a Gaussian wavepacket showing only the probability densities of the spatial and momentum state wave functions. Change the slider control for **Delta p** to five different values that span the range of the **Delta p** slider control. The program calculates the new probability distributions and repositions the **Delta x** control to the appropriate value. After each adjustment of the **Delta p** slider control, observe the effects of this control on the probability densities and record the value of the **Delta p** slider and the value of the **Delta x** slider. Calculate the product of **Delta p** and **Delta x** for each setting of the **Delta p** slider.

7.2   **Gaussian Wavepackets**

Run the program QUANTUM. The initial section is the uncertainty principle section of the program. a) Change the **Mean p** slider control to approximately 3. What changes do you observe in the probability densities in position and momentum space? b) Now press F6 to display the real and imaginary wavefunction components in addition to the probability densities. Change the **Mean k** slider control back to approximately zero. What changes do you observe in the real and imaginary parts of the position space wavefunction. c) Change the **Mean p** slider control back and forth between approximately +3 and -3 and observe the difference in the two spatial state wavefunctions for the two values of **Mean p**. Does the real part lead or lag the imaginary part of the wave? That is, are the maxima and minima of the real part of the wave slightly ahead of (leading) or slightly behind (lagging) the maxima and minima of the imaginary part of the wavefunction in the direction of the **Mean p**? d) Explain how your answer to part c is consistent with the Schrödinger equation.

7.3   **More Gaussian Wavepackets**

Repeat each part of Exercise 7.2 reversing the role of x and p.

7.4   **Limit of Gaussian Wavepacket**

Run the program QUANTUM. The program starts in the uncertainty principle section. Press the **F6-Show Real** hot key to display the real and imaginary components of the wavefunction. Move the **Delta x** slider to the maximum value allowed. Change the value of the **Mean p** slider to some value other than zero. What functions should the wavefunction and momentum distribution functions approach as **Delta x** becomes larger and larger? Do the functions displayed by the program appear to have the proper behavior?

7.5   **Other Wavepacket Shapes**

Start up the QUANTUM program. a) Select **Parameters—Wave function parameters**. On the **Wavefunction parameters** input screen select a triangular wave and accept the input screen. Change the **Delta x** slider to about five different values of **Delta x** that span the range of **Delta x**, recording **Delta x** and **Delta p** for each setting of the **Delta x** slider. b) Calculate the product of **Delta x** and **Delta p** for each of the recorded pairs of values. c) Select **Parameters—Wave function parameters** and this time select rectangular wave function and accept the input screen. Observe that for all values of **Delta x**, the value of **Delta p** is infinite. Why do more and more large momentum components need to be included as you go from Gaussian wavepackets, to triangular wavepackets, to rectangular wavepackets? (Consider the steepest slope of the wavepacket and the maximum slope of a component wave of momentum p.)

7.6   **Phase and Group Velocity**

Start up the QUANTUM program. Select **Section—Time-dependent free**

particles. The wavefunction should be a Gaussian distribution and the dispersion relation should be that resulting from Schrödinger's equation. The real and imaginary parts of the wavefunction as well as the probability density should be displayed. a) Determine the velocity of the phase of the wave by pressing the **F2-Run** hot key and measuring the distance a peak of either the real or imaginary part of the wavefunction advances on the screen in a given time. b) Determine the group velocity of the wavepacket by measuring the distance the peak of the wavepacket moves forward in a given time. c) Calculate the expected phase and group velocities based on Eqs. 7.33 and 7.35. d) Does the simulation result in the values predicted by theory?

7.7   Standing Waves

An interesting phenomena that is seen in the behavior of electron wavefunctions in a solid occurs when the dispersion relation reaches a relative minimum or maximum at a wave vector other than zero. This can occur because of the potential of the atoms within the solid. A wavepacket with a mean wavevector at the relative extremum will have zero group velocity even though none of the component eigenwaves has a zero wavevector. Select the cubic dispersion relation from the dispersion relations menu. This has a relative maximum at $k = 2.0$. Set the mean wave vector to 2.0 and observe the behavior of the motion of the wave packet and the phase velocity. Press the **F2-Run** hot key and allow time to step forward. a) Determine the phase velocity of the wave by measuring the distance a peak on either the real or imaginary parts of the wavefunction moves forward in time in a given time interval. b) Determine the group velocity by measuring how far the peak of the probability moves forward in a given time interval.

7.8   Phase and Group Velocities in Opposite Directions

Another interesting phenomena of wavefunctions in a solid material is in a region near a relative maximum of the dispersion relation. Select **Section—Time-independent free particles**. Select the dispersion to be cubic. Set the mean wave vector of the wavepacket to 2.1. Press the **F2-Run** hot key. a) Which direction does the wavepacket move? b) Which direction does the phase of the wave advance? c) Determine the phase velocity of the wave by measuring the distance a peak on either the real or imaginary parts of the wavefunction moves forward in time in a given time interval. d) Determine the group velocity by measuring how far the peak of the probability moves forward in a given time interval.

7.9   Finite Square Well

Start the program QUANTUM. Select **Section—Bound particles**. Select **Parameters—Well parameters**. Set a well depth of 100 eV and a well width of 1 nm. Solve for the bound state energy eigenfunctions and energy eigenvalues. How many bound states are there? Compare the energy eigenvalues to those of an infinite depth square well of the same width. Are the lower energy levels or higher energy levels shifted more? Why?

7.10   Unsymmetric Finite Square Well

Select **Section—Bound particles in a well**. Select **Parameters—Well parameters**. Input a square well with a different potential on the left side

of the well, 100 eV, than on the right side of the well, 50 eV, and a width of 1 nm. Solve for the energy eigenfunctions and energy eigenvalues. Solve for the energy eigenvalues for a symmetric 100 eV deep well and a symmetric 50 eV deep well. Are the eigenvalues of the unsymmetric well closer to the 100 eV deep symmetric well or closer to the 50 eV deep symmetric well? Why?

7.11   **Ramsauer Effect**

The interaction of free electrons with inert gas atoms can be modeled by an attractive square well potential with a width of the radius of the atom. Free electrons of some particular energies have very little interaction with the atoms, corresponding to a zero reflection coefficient in a one dimensional barrier problem. Suppose electrons of energy 1.0 eV show no interaction with an inert gas atom of radius 0.24 nm. Use the barriers section of the QUANTUM program to determine the depth of a square well with a width equal to the atomic radius that models this atom. What you need to do is to try different well depths of width 0.24 nm to find what depth gives zero reflection coefficient for electrons of energy 1.0 eV.

7.12   **Nonreflective Barriers**

In optics, nonreflective coatings can be made for specific wavelengths of light by putting a coating of a specific refractive index and specific thickness on the surface of optical elements. Using the barrier section of the program, verify that a similar result holds for quantum mechanics. a) Select **Parameters—Barrier parameters** and set up a potential with three regions with the third region having a potential of 1. For a wavefunction of energy 2, determine the width and potential of region 2 that must be used to ensure no reflection from the barrier.

7.13   **Diatomic Molecule Model**

Use the bound solutions section of the program to study the ground state as a function of separation of two wells of fixed width and depth. This is a simple model of a one-dimensional diatomic molecule. The "nuclei" of the molecule are being treated as stationary when the calculation of the "electron" state is calculated. It happens that this is a fairly good approximation in real molecules as the electron responds rapidly to nuclear motion. See exercise 7.14 for more about the motion of the nuclei. From the **Parameters** menu, select **Well parameters**. Select a user defined potential and define a double well by entering a function

$$2 * (-H(-L - 0.5) + H(-L + 0.5)$$
$$-H(L - 0.5) + H(L + 0.5)). \tag{7.67}$$

The parameter L is the separation of the center of two wells of width 1 and depth 2. When the value of the separation is less than 1, the wells overlap and the potential in the region of overlap is the sum of the two well potentials.

Starting with the wells a distance 2.0 apart as in the function given and solve for the ground state energy. Record the ground state energy as a function of position for each distance as the separation of the wells is decreased in steps of one-tenth down to zero. Plot the ground state energy as a function of the separation value. What do these results mean? Are these results expected? If the wells were free to move, what would they do? Is this the behavior of real diatomic atoms?

The above potential does not include any interaction between the wells themselves as would be the case with a real atomic molecule. In real atoms there is a repulsive potential between the nuclei. Add in a term that has the form of a repulsive Coulomb potential between the wells. This will just change the user define potential to be

$$2 * (-H(-L - 0.5) + H(-L + 0.5) - H(L - 0.5)$$
$$+ H(L + 0.5)) + 1/(L * L). \tag{7.68}$$

Again find the ground state energy for each value of L between 0 and 1 in 0.1 steps. Plot the ground state energy now as a function of separation L. Is this qualitatively the expected result for a diatomic molecule? What would you expect the mean bond length to be for this diatomic model?

7.14 Diatomic Molecule Vibration

The approximate vibrational spectrum of the model diatomic molecule in Exercise 7.13 can be determined from the data of that exercise. The ground state energy as a function of separation can be considered as an effective potential energy for the "nuclear" motion. After taking the data of Exercise 7.13, fit a parabola around the minimum of the ground state energy versus well separation curve. Do this by picking three points near the minimum and solving for the coefficients of a quadratic that fit the three points. Input a quadratic with this potential into the input screen when **Parameters—Well parameters** is selected. Solve for the energy eigenfunctions and energy eigenvalues for this potential.

7.15 Numerical Method Check

The single and double square well, barrier, and step simulations all numerically solve the exact solution to the time-independent Schrödinger equation. The solution to the user input potential section is a numerical solution that numerically integrates Shrodinger's equation. Verify the accuracy of the numerical solution by defining a user defined square well and comparing the energy eigenvalues to those of the exact solution for an identical well.

7.16 Time-Dependent Potential

In the introduction, it was stated that Schrödinger's equation is separable if the potential does not depend on time. It may be separable even if there is a time dependence to the potential. In particular, if

$$V(x, t) = V_1(x) + V_2(t). \tag{7.69}$$

Schrödinger's equation is still separable into

$$\frac{\hbar^2}{2m} \frac{d^2 X(x)}{dx^2} + V_1(x) X(x) = E X(x) \tag{7.70}$$

and

$$i\hbar \frac{dT(t)}{dt} - V_2(t)T(t) = ET(t). \tag{7.71}$$

Suppose

$$V_2(t) = V_0 \cos(\omega t). \tag{7.72}$$

Show that $T(t)$, the time dependence of the solutions to Schrödinger's equation, is given by

$$T(t) = e^{\frac{-i}{\hbar}Et} + \frac{A}{\omega} \sin(\omega t). \tag{7.73}$$

7.17 Minimal Uncertainty

Verify that, of the function shapes given in the time-independent free particle simulation, the Gaussian functions satisfy

$$\Delta p \Delta x = \frac{\hbar}{2} \tag{7.74}$$

independent of the width of the wavefunction, and the product is greater for all of the other function types. Define several of your own functions to see if there is a distribution for which the product of Δp and Δx is smaller. What characteristic of the spatial wavefunction determines the width of the distribution of momentum?

7.18 WKB Approximation

When the energy of the wavefunction is considerably less than the height of the barrier and the width of the barrier is much greater than the magnitude of the wavelength, the transmission through a barrier is given by the Wentzel-Kramers-Brillouin (WKB) approximation,

$$T = e^{-2 \int 2K dx}. \tag{7.75}$$

Start with a narrow barrier of height twice the energy of the incident waves and gradually increase the thickness of the barrier to verify that the transmission coefficient approaches the WKB approximation. Record values of the transmission coefficient calculated by the program and values calculated using the WKB approximation as a function of barrier thickness.

7.19 Bouncing Ball

A ball bouncing on a table has an effectively infinite potential for vertical positions below the surface of the table and a potential

$$V(x) = mgx \tag{7.76}$$

for positions above the table where the height above the table is x. Use the "define your own potential" section to solve for the energy eigenvalues for an object with the mass of an electron bouncing on a table.

7.20   Conservation of Probability

Conservation equations of a quantity can be written in a differential form
such that the time derivative of the density added to the divergence of the
flux is equal to zero. Because the probability density in quantum mechanics
is $\psi^*\psi$, conservation of probability density can be written as

$$\frac{\partial}{\partial t}(\psi^*\psi) + \frac{\partial}{\partial x}j = 0, \tag{7.77}$$

where j is the probability flux. Use Schrödinger's equation and the complex
conjugate of Schrödinger's equation to show

$$j = \frac{i\hbar}{2m}(\psi\frac{d}{dx}\psi^* - \psi^*\frac{d}{dx}\psi). \tag{7.78}$$

7.21   Probability Flux for Traveling Waves

For wavefunctions of the form

$$\psi = Ae^{i(kx-\omega t)} + Be^{-i(kx-\omega t)}, \tag{7.79}$$

show that the probability flux is given by

$$j = (A^*A - B^*B)\frac{\hbar k}{m} \tag{7.80}$$

when k is real.

7.22   Probability Flux of Evanescent Waves

For wavefunctions of the form

$$\psi = Ae^{i(kx-\omega t)} + Be^{-i(kx-\omega t)} \tag{7.81}$$

show that the probability flux is given by

$$j = (A^*B - AB^*)\frac{\hbar k}{m} \tag{7.82}$$

when k is imaginary.

7.23   Numerical Method Check

Whenever a numerical solution to a problem is found, one must be somewhat careful that the algorithm used gives a good approximation to the actual solutions. Making discrete steps when integrating Schrödinger's equation can introduce errors into the results. In order to check the numerical
method used in the **Bound Solutions** section of the program for user defined potentials, we can input a user defined square well and compare the
results for the exact square well solutions. First, find the Hamiltonian eigenvalues and eigenfunctions for an exact square well with a depth of 150 and
a width of 1. Examine the shapes of the Hamiltonian eigenfunctions and

record the eigenvalues for the exact solutions. Next, input a user defined potential

$$150 - 150 * (H(x) - H(x - 1))$$ (7.83)

giving a domain of 0,1 and selecting softwalls. This gives a square well identical to the one defined for which the exact solutions were found. Allow the program to find all of the Hamiltonian eigenvalues and eigenfunctions. Record the Hamiltonian eigenvalues and compare them to the ones found for the exact solution. Are these approximate solutions reasonable approximations?

7.24 Direct Numerical Integration of Schrödinger's Equation

It may appear tempting to integrate Schrödinger's time-dependent equation in the same way in which Schrödinger's time-independent equation is integrated by using a discrete approximation to the derivatives of the wavefunction as in section 7.2. Schrödinger's time-dependent equation becomes

$$\frac{-\hbar^2}{2m} \frac{\Psi(x_n, t) + \Psi(x_{n-2}, t) - 2\Psi(x_{n-1}, t)}{(\Delta x)^2} + V(x_{n-1})\Psi(x_{n-1}, t)$$

$$= i\hbar \frac{\Psi(x_{n-1}, t + \Delta t) - \Psi(x_{n-1}, t)}{\Delta t}$$ (7.84)

where Δt is the size of the time step taken.

7.25 Pertubation Theory

There are not a great number of potentials for which Schrödinger's equation can be solved exactly. This program finds numerical solutions by integrating Schrödinger's equation for all but the square well case. Another method that is most useful when the potential is almost equal to a potential for which the solution is known is provided by perturbation theory. The small difference between actual potential and the solvable potential is called the perturbing potential. Perturbation theory allows one to write the solutions of the actual time-independent Schrödinger's equation as a linear combination of the solutions to a solvable Schrödinger's equation. It also allows one to write the Hamiltonian eigenvalues as a series involving powers of the perturbing Hamiltonian.[5]

The first-order correction to the Hamiltonian eigenvalue of a given state is given by the expected value of the perturbing potential for the corresponding state of the solvable Hamiltonian. That is,

$$\Delta H_1(i) = <i|H'|i> = \int \psi^* H' \psi dx,$$ (7.85)

where $\Delta H_1(i)$ is the first-order change in the Hamiltonian eigenvalue of the state labeled i and H' is the perturbing potential.

Check whether this is a good approximation for the case of adding a small bump to a square well potential. In the **Bound Solutions** section of the program, first determine the Hamiltonian eigenvalues by selecting the exact solutions to a square well with a depth of 150 and a width of 1. Next, define a potential using the user defined potential option that is a square well with a depth of 150 and width of 1, but also add in a bump in the center of the well with a height of 30 and a width of 0.2. This can be done by entering the function

$$150 - 150 * (H(x) + H(x - 1))$$
$$+30 * (H(x + 0.1) - H(x + 0.1)) \qquad (7.86)$$

giving a domain of 0,1 and selecting softwalls. After the program has calculated the Hamiltonian eigenvalues and eigenfunctions, select a wavepacket that has a single component of the first eigenfunction. Use the integration tool to find the integral of the potential multiplied by the probability dsensity over the range of the bump. This is ΔH_1 given above. Find the difference between the first Hamiltonian eigenvalue for the bumped square well and the exact square well. Is ΔH_1 a reasonable approximation to this difference?

Repeat the calculation of ΔH_1 for the remaining Hamiltonian eigenfunctions for this potential.

7.26   More Perturbation Theory

To show that first order perturbation doesn't always give a good approximation to the shift in the Hamiltonian eigenvalues, repeat the previous exercise, but this time add a bump that is somewhat larger. For instance, you may try

$$V(x) = 150 - 150 * (H(x + 1) + H(x - 1))$$
$$+70 * (H(x + 0.3) - H(x + 0.3)) \qquad (7.87)$$

as the potential. Does the first-order correction to the Hamiltonian eigenvalues predict the difference between this bumped square well and the related exact square well Hamiltonian eigenvalues?

7.27   Relativistic Correction

Certainly Schrödinger's equation is not relativistically correct. The kinetic energy term in Schrödinger's equation, $p^2/2m$, is only the lowest order approximation to the relativistic kinetic energy

$$K_{relativistic} = \sqrt{p^2c^2 + m^2c^4} - mc^2. \qquad (7.88)$$

The next higher-order approximation to the relativistic energy is given by

$$K_{relativistic} \frac{p^2}{2m} + \frac{p^4}{8m^3c^4}. \qquad (7.89)$$

With this higher-order approximation to the relativistic kinetic energy, Schrödinger's equation becomes

$$-\frac{\hbar^2}{2m}\frac{d^2\psi(x)}{dx^2} - \frac{\hbar^4}{8m^3c^2}\frac{d^4\psi(x)}{dx^4} + V(x)\psi(x) = i\hbar\frac{\partial\psi(x)}{\partial t}. \qquad (7.90)$$

Verify that for a free particle ($V = 0$), plane wave solutions of the form

$$\Psi(x, t) = Ae^{i(kx - \omega t)} \tag{7.91}$$

are still solutions to this new wave equation with a dispersion relation

$$\omega(k) = \frac{\hbar^2 k^2}{2m} + \frac{\hbar^4 k^4}{8m^3 c^2}. \tag{7.92}$$

7.28   Circular Symmetric Potentials
In two dimensions, Schrödinger's equation separates into two equations in circular polar coordinates r and θ when the potential is a central a potential,

$$\psi(r, \theta) = R(r)\Theta(\theta). \tag{7.93}$$

In circular coordinates,

$$\nabla^2 \psi = \frac{\partial^2 \psi}{\partial r^2} + \frac{1}{r} \frac{\partial \psi}{\partial} + \frac{1}{r^2} \frac{\partial^2 \psi}{\partial \theta^2}. \tag{7.94}$$

a) Show that Schrödinger's time-independent equation separates into

$$\frac{\partial^2 \Theta(\theta)}{\partial \theta^2} + m^2 \Theta(\theta) = 0 \tag{7.95}$$

$$\frac{\partial^2 R(r)}{\partial r^2} + \frac{1}{r} \frac{\partial R(r)}{\partial r} + (\frac{-m^2}{r^2} + V(r) - E)R(r) = 0, \tag{7.96}$$

where $-m^2$ is the separation constant. The equation in θ is easily solved,

$$\Theta(\theta) = Ae^{im\theta} + Be^{-im\theta}. \tag{7.97}$$

For the solution to be single valued, $\Theta(\theta)$ must be equal to $\Theta(\theta + 2\pi)$. It is easy to see that this implies that m must be an integer. As the angular solutions have now been found, the solution of Schrödinger's time-independent equation has been reduced to a one-dimensional equation for the radial functions. The domain of the radial functions is zero to infinity. b) Start the program QUANTUM. Select **Bound Solutions** from the **Section** menu. From the **Parameters** menu select **Well Parameters**. Check the button for **Radial 2D Schrödinger equation**. This places the program in the mode to solve for the solutions of the radial equation above. Solve for the lowest energy eigenfunctions for a circular well potential

$$V(r) = 100 * H(r - 1) \tag{7.98}$$

and a domain of 0 to 1 for values of m equal to 0, 1, 2.

7.29   Two-Dimensional Atom
Start the QUANTUM program and place it in the mode to solve the

two-dimensional radial Schrödinger equation as in Exercise 7.29. A two-dimensional atom with one electron can be modeled by a potential

$$V(r) = -ke^2/r, \qquad (7.99)$$

which can be entered into the user input potential as

$$V(r) = -1.44/r. \qquad (7.100)$$

Set the domain equal to 0–10 nm. Solve for the energy eigenvalues for this circular atom.

7.30   Another Two-Dimensional Atom

Start the QUANTUM program and place it in the mode to solve the two-dimensional radial Schrödinger equation as in Exercise 7.29. Another two-dimensional atom could be modeled with a logarithmic potential,

$$V(r) = ke^2 ln(r), \qquad (7.101)$$

This would be the dependence of the potential on radial distance if a two-dimensional Gauss's law held true. Enter this into the user input potential as

$$V(r) = 1.44 ln(r). \qquad (7.102)$$

Set the domain equal to 0–10 nm and solve for the energy eigenvalues for this form of two-dimensional atom.

7.31   Spherically Symmetric Potentials

Although two- and three-dimensional problems are not directly solvable with this program, many of the interesting cases in the higher dimensions are separable into one-dimensional problems. In particular, central potential problems (problems in which the potential only depends on the radial distance from the center of potential) are separable. This is covered in most modern physics texts in preparation for solving the one-electron atom problem.[3] Of course, the one-dimensional equation that results from the separation of variables is not identical to Schrödinger's equation in one dimension, but looks somewhat similar. In three dimensions when the potential is spherically symmetric, Schrödinger's equation separates into three equations in spherical polar position coordinates r, θ, and phi. These equations are

$$\frac{\partial^2 \Theta(\theta)}{\partial \theta^2} = m\Theta(\theta) \qquad (7.103)$$

$$\frac{1}{\sin \theta} \frac{\partial^2 \Theta(\theta)}{\partial \theta^2} = C\Theta(\theta) \qquad (7.104)$$

$$\frac{\partial^2 R(r)}{\partial r^2} + \frac{2}{r} \frac{\partial R(r)}{\partial r} + (\frac{l(l+1)}{r^2} + V(r))R(r) = ER(r) \qquad (7.105)$$

where $\psi(r, \theta, \phi) = R(r)\Theta(\theta)\Phi(\phi)$.

The constants m and l are separation constants. Notice that the first equation in coordinate ϕ and the second equation in coordinate θ have no dependence on the potential. Both equations have solutions that don't depend on the form of the spherically symmetric potential. The solutions are well known and are discussed in many modern physics texts and almost all quantum mechanics texts. Start the QUANTUM program and select **Section—Bound states**. Select **Parameters—Well parameters**. On the input screen click on the three-dimensional radial for type of wave equation. For the potential, input

$$V(r) = 100 * H(1 - r) \tag{7.106}$$

and a domain of 0–1. Solve for the eigenvalues for this potential.

References

1. Ohanian, H.C. *Principle of Quantum Mechanics*. Englewood Cliffs, NJ: Prentice Hall, 1990.

2. Rudin, W. *Real and Complex Analysis*. New York: McGraw-Hill, 1974.

3. Eisberg, R., Resnick, R. *Quantum Physics of Atoms, Molecules, Solids, Nuclei, and Particles*. New York: John Wiley & Sons, 1974.

4. Hiller, J.R., Johnston, I.D., Styer, D.F. *Quantum Mechanics Simulations*. New York: John Wiley & Sons, 1995.

5. Gasiorowicz, S. *Quantum Physics*. New York: John Wiley & Sons, 1974.

6. Schiff, L. I. *Quantum Mechanics*. New York: McGraw-Hill, 1968.

8

Hydrogenic Atoms and the H_2^+ Molecule

John R. Hiller

8.1   Introduction

Systems with a single light particle can be reasonably simple to analyze. Any heavy particles present can be treated as relatively slow moving, and become only fixed centers of repulsion or attraction. Such systems include hydrogen, hydrogenic ions, and diatomic molecules, all of which are cylindrically symmetric, if not spherically symmetric. The symmetry can be exploited to reduce the difficulty of calculations. Even when exposed to uniform electric and magnetic fields, they all may possess cylindrical symmetry.

In the case of hydrogen and hydrogenic atoms, when the internal interactions are modeled by a simple Coulomb potential, there are, of course, exact solutions. These solutions are briefly reviewed below.

When such atoms are exposed to electric and magnetic fields, the wave functions are distorted. Exposure to magnetic fields is of particular interest in astrophysics and condensed-matter physics.[1,2] The magnetic interactions are of three types: a coupling to the intrinsic magnetic moment associated with spin, a linear coupling to the orbital magnetic moment, and a quadratic term that, for a uniform field, can be written as an oscillator potential in the directions transverse to the field. We will include only the last of these in the numerical calculations. Spin is neglected entirely, and the linear term only results in an energy shift, with no change in the wave function. The potential associated with the electric interaction will also be treated numerically; if the field is uniform, then the interaction is linear in the coordinate along the field direction.

Diatomic molecules can be usefully treated in the Born-Oppenheimer approximation,[3] in which the motion of the electrons takes place in a background field of fixed nuclei. An effective potential energy for the nuclei is then obtained from combining their direct electrostatic repulsion with the binding energy of the electrons. Here we consider the case of single-electron diatomic ions. The Schrödinger equation for the electron is solved numerically[*] for a series of nuclear separations and the effective potential extracted. The Schrödinger equation for the relative motion of the nuclei is then solved, also numerically; this equation can be reduced to a simple radial equation that is easily solved via finite differences and ordinary matrix diagonalization.

Except for unperturbed hydrogenic ions, the Schrödinger equation for the electron reduces no further than a two-dimensional, cylindrical wave equation. To this equation we apply two numerical methods. The primary one is a finite-difference scheme combined with Lanczos diagonalization[5] of the resulting matrix; the Lanczos method is used because the matrix is quite large. The secondary method is a basis-function method that works well for weak electric and magnetic fields; the wave function is expanded in terms of hydrogenic eigenfunctions, and the coefficients of the expansion and the energies are obtained from matrix diagonalization.

These physical situations are simulated by the program HATOM. Most parameters are adjustable, and in some cases, such as field strengths, sequences of values can be automatically selected and the results recorded for playback in an animated form. The choice of "light" particle is also selectable by the user; thus the program can be used to study the effects of mass on molecular binding.

The following sections provide an introduction to the analysis of single-electron systems and to the use of the program.

8.2 *Atoms and Molecules With One Electron*

8.2.1 Hydrogen

The simplest atomic system is hydrogen. We generalize this case a little to include any single-electron ion that is well modeled by a purely Coulombic potential energy. The Schrödinger equation is[3,6–8]

$$-\frac{\hbar^2}{2\mu}\nabla^2\Psi(\mathbf{r}) - \frac{Ze^2}{r}\Psi(\mathbf{r}) = E\Psi(\mathbf{r}),\tag{8.1}$$

where μ is the reduced mass, $-e$ the electron charge, and Ze the nuclear charge. The reduced mass is obtained from the electron mass m_e and the nuclear mass m_N as

$$\mu = \frac{m_e m_N}{m_e + m_N}.\tag{8.2}$$

The charges are measured in Gaussian units.

Spherical coordinates (r, θ, ϕ) form the natural choice for this situation. Separation of variables then leads to two equations

$$\frac{1}{\sin\theta}\frac{\partial}{\partial\theta}\left(\sin\theta\frac{\partial Y(\theta,\phi)}{\partial\theta}\right) + \frac{1}{\sin^2\theta}\frac{\partial^2 Y(\theta,\phi)}{\partial\phi^2} = -l(l+1)Y(\theta,\phi),\tag{8.3}$$

---

[*]The electron problem can also be solved in a nearly analytic fashion by separation in confocal elliptic coordinates.[4]

and

$$-\frac{\hbar^2}{2\mu r^2}\frac{d}{dr}\left(r^2\frac{dR(r)}{dr}\right) + \frac{l(l+1)\hbar^2}{2\mu r^2}R(r) - \frac{Ze^2}{r}R(r) = ER(r), \tag{8.4}$$

with $\Psi(\mathbf{r}) = R(r)Y(\theta,\phi)$. The first has the spherical harmonics Y_{lm} as its solutions; these are given by

$$Y_{lm} = (-1)^m\sqrt{\frac{(2l+1)(l-m)!}{4\pi(l+m)!}}e^{im\phi}P_l^m(\cos\theta), \quad m = -l,\ldots,l, \tag{8.5}$$

where $P_l^m(\cos\theta)$ is an associated Legendre function. They are easily computed via a recursion relation[9]

$$P_{l+1}^m(x) = \frac{1}{l-m+1}[(2l+1)xP_l^m(x) - (l+m)P_{l-1}^m(x)], \tag{8.6}$$

and a convenient starting point

$$P_{m-1}^m(x) = 0, \quad P_m^m(x) = \frac{(2m)!}{2^m m!}(1-x^2)^{m/2}. \tag{8.7}$$

The second equation, Eq. 8.4, has solutions R_{nl} that can be written[10] in terms of Laguerre polynomials L_{n-l-1}^{2l+1}

$$R_{nl} = \sqrt{\left(\frac{2Z}{na_0}\right)^3\frac{(n-l-1)!}{2n[(n+l)!]^3}}e^{-\frac{z}{na_0}r}\left(\frac{2Z}{na_0}r\right)^l L_{n-l-1}^{2l+1}\left(\frac{2Z}{na_0}r\right),$$
$$l = 0,\ldots,n-1; \quad n = 1,2,\ldots, \tag{8.8}$$

with $a_0 = \hbar^2/\mu e^2$. The eigenenergies are

$$E_n = -\frac{Z^2}{n^2}\frac{e^2}{2a_0}; \tag{8.9}$$

for ordinary hydrogen, $e^2/2a_0$ is approximately 13.6 eV. The Laguerre polynomials can be computed from a recursion relation,[9]

$$L_{k+1}^\alpha(t) = \frac{1}{k+1}[(2k+\alpha+1-t)L_k^\alpha(t) - (k+\alpha)L_{k-1}^\alpha(t)], \tag{8.10}$$

and the two lowest-order forms:

$$L_0^\alpha(t) = 1, \quad L_1^\alpha(t) = 1 + \alpha - t. \tag{8.11}$$

We then have the means by which to compute the hydrogen eigenfunctions and energies for any choice of n, l, m, Z, and μ. The program HATOM incorporates these and can be used to obtain plots of the probability density $|\Psi|^2 \propto (R_{nl}P_l^m)^2$.

8.2.2   Hydrogenic Atoms in Uniform Fields

When interactions with external fields are added to the hydrogenic Schrödinger equation given in Eq. 8.1, we have[3,6]

$$-\frac{\hbar^2}{2\mu}\nabla^2\Psi - \frac{Ze^2}{r}\Psi - \frac{e}{2\mu}\mathbf{B}\cdot\mathbf{L}\Psi + \frac{e^2}{2\mu}A^2\Psi - e\Phi\Psi = E\Psi, \qquad (8.12)$$

with $\mathbf{L}$ the angular momentum, $\mathbf{B} = \nabla \times \mathbf{A}$ the magnetic field, $\mathbf{A}$ the vector potential, and Φ the electrostatic potential.[*] The electrostatic field $\mathbf{E}$ is given by $-\nabla\Phi$. The magnetic interaction with the spin magnetic moment has been excluded. An interesting scale for the magnetic field is a strength on the order of 1 MT (10^6 Tesla), for which the quadratic magnetic term is of importance equal to that of the Coulomb term.[1] A typical scale for the electric field is 0.1 TV/m (10^{11} volts per meter).

We will consider the case where the fields are uniform and point in the same direction. The common direction is taken to define the z-axis. The linear magnetic term becomes $-(e/2\mu)BL_z\Psi$, which will be seen to contribute only a constant to the energy. The vector potential for a uniform field can be written as $\mathbf{A} = \frac{1}{2}\mathbf{B} \times \mathbf{r}$, and its square reduces to $A^2 = \frac{1}{4}B^2(x^2 + y^2)$. The electrostatic potential may be chosen to be $\Phi = -E_z z$.

The potential energy terms are naturally expressed in terms of cylindrical coordinates (ρ, ϕ, z). The angular momentum operator L_z is a differential operator $-i\hbar(\partial/\partial\phi)$. The other two terms depend explicitly on $\rho = \sqrt{x^2 + y^2}$ and z. In these coordinates the Schrödinger equation (Eq. 8.12) becomes

$$-\frac{\hbar^2}{2\mu}\left(\frac{\partial^2\Psi}{\partial\rho^2} + \frac{1}{\rho}\frac{\partial\Psi}{\partial\rho} + \frac{1}{\rho^2}\frac{\partial^2\Psi}{\partial\phi^2} + \frac{\partial^2\Psi}{\partial z^2}\right) - \frac{Ze^2}{r}\Psi$$

$$+i\hbar\frac{eB}{2\mu}\frac{\partial\Psi}{\partial\phi} + eE_z z\Psi = E\Psi. \qquad (8.13)$$

For this equation, we seek solutions of the form $\Psi = e^{im\phi}\psi_m(\rho, z)/\sqrt{2\pi}$. The function ψ_m must satisfy

$$-\frac{\hbar^2}{2\mu}\left(\frac{\partial^2\psi_m}{\partial\rho^2} + \frac{1}{\rho}\frac{\partial\psi_m}{\partial\rho} + \frac{\partial^2\psi_m}{\partial z^2}\right)$$

$$+ \left[\frac{\hbar^2 m^2}{2\mu\rho^2} - \frac{Ze^2}{\sqrt{\rho^2 + z^2}} + \frac{e^2 B^2}{8\mu}\rho^2 + eE_z z\right]\psi_m$$

$$- \frac{eB}{2\mu}\hbar m\psi_m = E\psi_m. \qquad (8.14)$$

The linear magnetic term is now clearly seen as a simple energy shift of magnitude $(e/2\mu)B\hbar m$; the program HATOM neglects it. The remaining magnetic interaction

---

[*]For a more careful treatment of the two-body problem in a magnetic field, see reference 11.

is known as the quadratic Zeeman term.[12,13] The electric interaction drives what is known as the Stark effect.[12] Both shift the energy levels away from the standard Coulombic spectrum of a hydrogen atom.

For the electric interaction, one technicality must be addressed. As written in Eq. 8.14, bound states do not exist. The electrostatic potential is unbounded from below, and for any energy there will be a solution with finite probability of escape to infinity. In the usual perturbative treatment of the Stark effect, this is not a problem, but in our nonperturbative treatment it is, at least in principle. To remove the difficulty we will assume that the electric field extends for only a finite distance.

The presence of the external fields will distort the probability density of the electron. One measure of this distortion is the dipole moment,[3,8]

$$\mathbf{p} = \int d^3 r \, \Psi^*(-e\mathbf{r})\Psi \,. \tag{8.15}$$

The program HATOM computes and displays the z-component.

8.2.3   Single-Electron Diatomic Molecules

The binding of a diatomic molecule comes about because electrons are shared in such a way as to reduce the total energy below that of two separated atoms. The rearrangement of electron orbits takes place quite rapidly compared to motions of the nuclei; at any given nuclear separation one can then compute the eigenenergies of the electrons and the direct Coulomb repulsion of the nuclei to obtain an effective potential energy for the nuclei alone.

Let s be the separation of the nuclei. We place one nucleus, of charge $Z_1 e$, at $-\frac{1}{2}s\hat{z}$ and the other, of charge $Z_2 e$, at $+\frac{1}{2}s\hat{z}$. With s fixed, the Schrödinger equation for a single electron is[3,8]

$$-\frac{\hbar^2}{2m_e}\nabla^2\Psi - \frac{Z_1 e^2}{|\mathbf{r} + \frac{1}{2}s\hat{z}|}\Psi - \frac{Z_2 e^2}{|\mathbf{r} - \frac{1}{2}s\hat{z}|}\Psi = E(s)\Psi \,. \tag{8.16}$$

The eigenenergy E depends parametrically on s. When s is very large, the electron will be bound to one nucleus. The lowest possible energy is obtained when it is bound to the nucleus with the larger charge $Z_>$; this yields $E(\infty) = -(e^2/2a_0)Z_>^2$. It is relative to this energy that the effective nuclear potential is measured. The effective potential then takes the form[14]

$$V_{\text{eff}}(s) = \frac{Z_1 Z_2 e^2}{s} + E(s) - E(\infty) \,. \tag{8.17}$$

Once the nuclear potential is calculated, the binding energy of the molecule can be estimated. The relative motion of the nuclei obeys the radial Schrödinger equation,

$$-\frac{\hbar^2}{2\mu_N s^2}\frac{d}{ds}\left(s^2\frac{dR_l(s)}{ds}\right) + \left[V_{\text{eff}}(s) + \frac{l(l+1)\hbar^2}{2\mu_N s^2}\right]R_l(s) = E_l R_l(s) \,, \tag{8.18}$$

where $\mu_N = m_1 m_2/(m_1 + m_2)$ is the reduced nuclear mass; and m_1 and m_2, the individual nuclear masses. The complete wave function is, of course, $R_l(s)Y_{lm}(\theta, \phi)$.

The program HATOM solves this binding problem numerically in two steps. One first directs the program to solve the electron problem in the presence of the two charges, and to record the effective potential, for an automated sequence of separations. The electron wave function is plotted at each separation and can be recorded for more rapid viewing in a playback mode. Besides the automated calculation, there are other options that allow further exploration of the electron wave function.

Once V_{eff} is recorded, the computation of the binding of the molecule may be selected. The radial Schrödinger equation (Eq. 8.18) is then solved, and the results are displayed in much the same way as the analytic solutions for the undistorted hydrogen atom. The number of nodes in the radial wave function is controlled by the quantum number n_r, which is related to the principal quantum number n by $n = n_r + l + 1$. Selection of n_r fixes the vibrational levels, just as selection of l determines the rotational levels.[7]

8.3 Computational Approach

The two types of Schrödinger equations that must be solved are the cylindrical wave equation

$$-\frac{\hbar^2}{2\mu}\left(\frac{\partial^2\psi_m}{\partial\rho^2} + \frac{1}{\rho}\frac{\partial\psi_m}{\partial\rho} + \frac{\partial^2\psi_m}{\partial z^2}\right) + \left[V(\rho,z) + \frac{\hbar^2 m^2}{2\mu\rho^2}\right]\psi_m = E\psi_m, \quad (8.19)$$

and the radial wave equation

$$-\frac{\hbar^2}{2\mu_N r^2}\frac{d}{dr}\left(r^2\frac{dR_l}{dr}\right) + \left[V_{eff}(r) + \frac{l(l+1)\hbar^2}{2\mu_N r^2}\right]R_l = E_l R_l. \quad (8.20)$$

The first will be approximated in two ways, by a finite-difference approximation and by a basis-function expansion; both methods require diagonalization of a matrix. Because the solutions are only approximate, methods for estimating the dipole moment, defined in Eq. 8.15, must also be discussed. The second wave equation (Eq. 8.20) will also be approximated by a finite-difference equation; for this second-order ordinary differential equation the matrix obtained is tridiagonal and, therefore, relatively easy to diagonalize.

8.3.1 Scaled Variables

Almost all of the calculations internal to the program HATOM are done in terms of scaled variables. The potential energy and the eigenvalues are computed relative to a natural energy scale $V_0 = \mu e^4/2\hbar^2$. The scaled energies are written as

$$\overline{V} = V/V_0 \text{ and } \overline{E} = E/V_0. \quad (8.21)$$

Distances are measured relative to a natural length scale $L_0 = \hbar^2/\mu e^2$. We define

$$\overline{r} = r/L_0, \quad \overline{\rho} = \rho/L_0, \quad \overline{z} = Z/L_0. \quad (8.22)$$

If μ is the electron mass, then V_0 is ~ 13.6 eV, and L_0 is the Bohr radius. In the case of the radial equation for nuclei, the energy and length scales are computed from the mass of the *light* particle, not the masses of the nuclei; this yields the appropriate molecular scales.

The scaled forms of Eqs. 8.19 and 8.20 are

$$-\left(\frac{\partial^2 \psi_m}{\partial \bar\rho^2} + \frac{1}{\bar\rho}\frac{\partial \psi_m}{\partial \bar\rho} + \frac{\partial^2 \psi_m}{\partial \bar z^2}\right) + \left[\zeta \overline{V}(\bar\rho, \bar z) + \frac{m^2}{\bar\rho^2}\right]\psi_m = \zeta \overline{E}\psi_m, \qquad (8.23)$$

and

$$-\frac{1}{\bar r^2}\frac{d}{d\bar r}\left(\bar r^2 \frac{dR_l}{d\bar r}\right) + \left[\zeta_N \overline{V}_{\text{eff}}(\bar r) + \frac{l(l+1)}{\bar r^2}\right]R_l = \zeta_N \overline{E}_l R_l, \qquad (8.24)$$

where $\zeta \equiv 2\mu V_0 L_0^2/\hbar^2$ and $\zeta_N \equiv 2\mu_N V_0 L_0^2/\hbar^2$ are dimensionless parameters. These parameters are the only ways in which the masses enter the numerical analysis in the remaining sections.

8.3.2   Finite-Difference Method

To approximate the cylindrical wave equation (Eq. 8.23) we construct a grid of points $(\bar\rho_j, \bar z_i)$ and replace the derivatives of ψ_m by finite differences.[15] In order that the grid be finite in size, we assume that ψ_m is zero beyond some finite range $\bar\rho_{\max}$ in $\bar\rho$ and $\bar z_{\max}$ in $\bar z$. To avoid singularities along $\bar\rho = 0$, the grid points are chosen with an offset. Let N_ρ and N_z be the number of points in the $\bar\rho$ direction and the positive $\bar z$ direction, respectively, and define grid spacings by

$$\Delta\rho = \frac{\bar\rho_{\max}}{N_\rho - 1/2}, \quad \Delta z = \frac{\bar z_{\max}}{N_z}. \qquad (8.25)$$

The grid points are then given by

$$\bar\rho_j = (j - 1/2)\Delta\rho, \quad j = 1,\dots,N_\rho$$
$$\bar z = i\Delta z, \quad i = -N_z,\dots,0,\dots,N_z. \qquad (8.26)$$

The values of functions at the grid points are indicated by subscripts. For example, we use $\psi_m(\bar\rho_j, \bar z_i) = \psi_{m,ji}$ and $V(\bar\rho_j, \bar z_i) = V_{ji}$. If an electric field contributes to V, it is assumed to be non-zero for at most a finite distance outside the grid

The finite-difference representation for the cylindrical wave equation is, then

$$-\frac{\psi_{m,j+1,i} - 2\psi_{m,ji} + \psi_{m,j-1,i}}{(\Delta\rho)^2} - \frac{1}{\bar\rho_j}\frac{\psi_{m,j+1,i} - \psi_{m,j-1,i}}{2\Delta\rho}$$
$$-\frac{\psi_{m,j,i+1} - 2\psi_{m,ji} + \psi_{m,j,i-1}}{(\Delta z)^2} + \frac{m^2}{\bar\rho_j^2}\psi_{m,ji} + \zeta \overline{V}_{ji}\psi_{m,ji} = \zeta \overline{E}\psi_{m,ji},$$
$$j = 1,\dots,N_\rho - 1, \quad i = -N_z + 1,\dots,N_z - 1. \qquad (8.27)$$

Besides direct substitution of finite-difference approximations, this can be obtained by application of a variational principle. The integrals in the expectation value of the energy operator are approximated by the rectangle rule[*] and the expectation value is minimized with respect to the values of $\psi_{m,ji}$, subject to the constraint of fixed normalization for ψ_m. The normalization integral is approximated in the same way as those in the expectation value. The eigenenergy $\overline{E}_m$ enters as a Lagrange multiplier. This alternate derivation makes clear that $\psi_{m,0i}$ is to be replaced by $\psi_{m,1i}$ where needed, in order to reproduce the correct variational equation for $j = 1$; otherwise, this replacement must be less strongly argued on the basis of azimuthal symmetry.

The finite-difference equations form a matrix eigenvalue problem. The values $\psi_{m,ji}$ are components of a vector indexed by the pair (j, i). Let H represent the matrix and $\mathbf{c}_m$ the vector, then the eigenvalue problem can be written as

$$H\mathbf{c}_m = \zeta\overline{E}_m\mathbf{c}_m . \tag{8.28}$$

The size of the matrix is typically quite large; it might be made of 10^6 elements or more. Diagonalization methods that require storage and direct manipulation of the matrix are then difficult to implement without using large amounts of memory.

As an alternative method that does not require storage of the full matrix, we use the Lanczos algorithm.[5] This algorithm is an iterative method that begins with an initial guess for an eigenvector and generates a sequence of orthogonal vectors that form a basis for a subspace in which the matrix H is tridiagonal. Completion of the diagonalization can then be done by more standard means.[16] Any symmetries of the initial vector that are conserved by H will be shared by all elements of the basis; thus, in general, not all eigenvectors and eigenvalues will be found from one initial vector. More damaging to the method is that round-off error will eventually destroy orthogonality as the iterations proceed.[17]

The advantage in the Lanczos method comes from the fact that a complete sequence of vectors is not needed. If generation of the sequence is halted after a small number of iterations, and the resulting tridiagonal matrix is diagonalized, one finds that the extreme eigenvalues of the original matrix are well approximated.[5] Therefore, the Lanczos algorithm can be easily used to obtain the lowest eigenenergy, for a given value of m, which is all that HATOM has been designed to attempt.

The sequence of basis vectors is constructed as follows. Given an initial vector $\mathbf{c}_m^{(1)}$, normalized to 1, compute

$$a_n = \mathbf{c}_m^{(n)T}H\mathbf{c}_m^{(n)},$$
$$\mathbf{c}_m^{(n+1)} = \frac{1}{b_{n+1}}[H\mathbf{c}_m^{(n)} - a_n\mathbf{c}_m^{(n)} - b_n\mathbf{c}_m^{(n-1)}], \tag{8.29}$$

where the value of b_{n+1} is chosen to normalize $\mathbf{c}_m^{(n+1)}$, except that b_1 is zero. The norm $\mathbf{c}_m^{(n)T}\mathbf{c}_m^{(n)}$ and the expectation value a_n must be computed with respect to the correct infinitesimal area element, which in cylindrical coordinates includes a factor of $\bar{\rho}$. For example, the norm is given by

$$\mathbf{c}_m^{(n)T}\mathbf{c}_m^{(n)} = \sum_{ij}\overline{\rho}_j\psi_{m,ji}^{(n)2} . \tag{8.30}$$

---

[*]The rectangle rule for a one-dimensional integral $\int_a^{a+h} f(x)dx$ is simply the use of the approximation $hf(a + h/2)$.

The elements of the tridiagonal matrix are the a_n, on the diagonal, and the b_n on the first codiagonal.

For purposes of efficiency, the basic algorithm in Eq. 8.29 is rearranged[18] to minimize multiplications by H and to minimize storage. For the latter, only the two most recent vectors are kept in RAM and all others are stored on disk until needed, one at a time, for reconstruction of the approximate eigenvector.

The initial vectors used in HATOM are built from the analytic solutions for hydrogenic ions. The solution with the lowest energy for a given magnetic quantum number m is used. When two charge centers are present, the program uses a symmetric combination of the solution for each charge center.

8.3.3   Basis-Function Method

For a hydrogenic ion in weak electric and magnetic fields, an eigenfunction will be only slightly changed from a purely Coulombic wave function. It then makes sense to approximate the eigenfunction by an expansion in a basis of Coulombic wave functions.[19] The coefficients in the expansion should be small except for the one that multiplies the closely related wave function. To obtain the coefficients, we require that the projection of the Schrödinger equation onto each chosen basis function vanish. This condition leads to a matrix eigenvalue problem.

Let the eigenfunction ψ_m be approximated by the expansion

$$\psi_m \simeq \sum_{n=1}^{n_{max}} \sum_{l=0}^{l_n} c_{nl} R_{nl} Y_{lm} , \tag{8.31}$$

with l_n the smaller of $n - 1$ and an absolute limit l_{max}. Substitution into the Schrödinger equation Eq. 8.14 yields

$$\sum_{n=1}^{n_{max}} \sum_{l=0}^{l_n} \left[E_n R_{nl} Y_{lm} + \frac{e^2 B^2}{8\mu} r^2 \sin^2 \theta R_{nl} Y_{lm} + e E_z r \cos \theta R_{nl} Y_{lm} \right] c_{nl}$$

$$\simeq E \sum_{n=1}^{n_{max}} \sum_{l=0}^{l_n} c_{nl} R_{nl} Y_{lm}, \tag{8.32}$$

where E_n is given in Eq. 8.9 and the linear magnetic term has been neglected. Projection of this equation is accomplished by integrating over all space the product with $R_{n'l'} Y_{l'm}^*$. We then obtain

$$E_{n'} c_{n'l'} + \sum_{n=1}^{n_{max}} \sum_{l=0}^{l_n} \left(\frac{e^2 B^2}{8\mu} I_{n'l',nl}^m + e E_z J_{n'l',nl}^m \right) c_{nl} \simeq E c_{n'l'} , \tag{8.33}$$

with

$$I_{n'l',nl}^m \equiv \int d^3 r R_{n'l'} Y_{l'm}^* r^2 \sin^2 \theta R_{nl} Y_{lm} \tag{8.34}$$

and

$$J_{n'l',nl}^m \equiv \int d^3 r R_{n'l'} Y_{l'm}^* r \cos \theta R_{nl} Y_{lm} . \tag{8.35}$$

The exponential decrease in the radial wave functions R_{nl} provides a natural cutoff for the contribution from the linear electric field.

We again have a matrix eigenvalue problem. The coefficients c_{nl} form a vector $\mathbf{c}$ and the left-hand side of Eq. 8.33 is the result of multiplication by a matrix A, with elements given by

$$A_{n'l',nl} = E_{n'}\delta_{n'n}\delta_{l'l} + \frac{e^2 B^2}{8\mu}I^m_{n'l',nl} + eE_z J^m_{n'l',nl}. \tag{8.36}$$

If n_{max} and l_{max} are kept small, as is sufficient for weak fields, the matrix is small enough for diagonalization by standard means.[16]

To complete the construction of the matrix A, the I and J integrals must be computed. The angular parts are easily done, after the following identities[10] are applied:

$$\cos\theta Y_{lm} = \sqrt{\frac{(l+m+1)(l-m+1)}{(2l+1)(2l+3)}}Y_{l+1,m}$$

$$+ \sqrt{\frac{(l+m)(l-m)}{(2l+1)(2l-1)}}Y_{l-1,m}$$

$$\sin\theta Y_{lm} = -\sqrt{\frac{(l+m+1)(l+m+2)}{(2l+1)(2l+3)}}e^{-i\phi}Y_{l+1,m+1}$$

$$+ \sqrt{\frac{(l-m)(l-m-1)}{(2l+1)(2l-1)}}e^{-i\phi}Y_{l-1,m+1}. \tag{8.37}$$

The angular integrals are

$$\int d\Omega Y^*_{l'm}\cos\theta Y_{lm} = \sqrt{\frac{(l+m+1)(l-m+1)}{(2l+1)(2l+3)}}\delta_{l',l+1}$$

$$+ \sqrt{\frac{(l+m)(l-m)}{(2l+1)(2l-1)}}\delta_{l',l-1} \tag{8.38}$$

and

$$\int d\Omega(\sin\theta Y_{l'm})^*\sin\theta Y_{lm}$$

$$= \left[\frac{(l+m+1)(l+m+2)}{(2l+1)(2l+3)} + \frac{(l-m)(l-m-1)}{(2l+1)(2l-1)}\right]\delta_{l'l}$$

$$- \sqrt{\frac{(l+m-1)(l+m)(l-m)(l-m-1)}{(2l-3)(2l-1)^2(2l+1)}}\delta_{l',l-2}$$

$$- \sqrt{\frac{(l-m+2)(l-m+1)(l+m+1)(l+m+2)}{(2l+5)(2l+3)^2(2l+1)}}\delta_{l',l+2}. \tag{8.39}$$

Because of the restrictions on l' obtained from the angular integrals, the needed radial integrals are few. Given the expression in Eq. 8.8 for the radial wave functions in terms of Laguerre polynomials L_k^α, and the expansion[9]

$$L_k^\alpha(t) = \sum_{j=0}^{k} (-1)^j \frac{(k+\alpha)!\, t^j}{(k-j)!(\alpha+j)!j!}, \qquad (8.40)$$

these integrals can be reduced to finite sums of ratios of factorials. We find

$$\int r^{s+2} dr\, R_{n'l'} R_{nl}$$

$$= \sqrt{\left(\frac{2Z}{n'a_0}\right)^3 \frac{(n'-l'-1)!}{2n'[(n'+l')!]^3} \left(\frac{2Z}{na_0}\right)^3 \frac{(n-l-1)!}{2n[(n+l)!]^3}}$$

$$\times \left(\frac{2Z}{n'a_0}\right)^{l'} \left(\frac{2Z}{na_0}\right)^l \sum_{j'=0}^{2l'+1} (-1)^{j'} \frac{(n'+l')!}{(n'-l-1-j)!(2l'+1+j')!j'!}$$

$$\times \sum_{j=0}^{2l+1} (-1)^j \frac{(n+l)!}{(n-l-1-j)!(2l+1+j)!j!}$$

$$\times \left(\frac{2Z}{n'a_0}\right)^{j'} \left(\frac{2Z}{na_0}\right)^j \frac{(l+l'+j+j'+s+2)!}{(\frac{Z}{n'a_0}+\frac{Z}{na_0})^{l+l'+j+j'+s+3}}. \qquad (8.41)$$

8.3.4 Estimation of the Dipole Moment

The dipole moment, defined in Eq. 8.15, is easily estimated, once an approximate wave function is obtained. The program HATOM stores the wave function as values at the grid points used in the finite-difference method. The integral that defines the dipole moment can then be approximated by the rectangle rule. We consider only the z-component and obtain

$$p_z \simeq -eL_0 \sum_{i=-N_z}^{N_z} \sum_{j=1}^{N_\rho} \bar\rho_j \bar z_i |\psi_{m,ji}|^2. \qquad (8.42)$$

The presence of $\bar\rho_j$ is dictated by the Jacobian for cylindrical coordinates.

When the basis-function method is used to obtain ψ_m, the wave function is again recorded by the program as values at the grid points only. The dipole moment is then calculated in the same way as for the finite-difference method. However, an alternative formula for p_z could be used. It is

$$p_z \simeq -e \sum_{n,n'=1}^{n_{\max}} \sum_{l,l'=0}^{l_n, l_{n'}} c_{n'l'} J_{n'l'nl}^m c_{nl}, \qquad (8.43)$$

where $J_{n'l',nl}^m$ is defined in Eq. 8.35. This alternative uses the coefficients of the basis-function expansion directly.

8.3.5 Radial Schrödinger Equation

To solve the scaled radial equation given in Eq. 8.24, we first replace the radial wave function R_l by $\bar{r}^{-1} u_l(\bar{r})$. The equation for u_l is

$$-\frac{d^2 u_l}{d\bar{r}^2} + \left[\zeta_N \bar{V}_{\text{eff}}(\bar{r}) + \frac{l(l+1)}{\bar{r}^2} \right] u_l = \zeta_N \bar{E}_l u_l, \tag{8.44}$$

which has the same structure as a one-dimensional Schrödinger equation. The appropriate boundary conditions are that u_l be zero at the origin and at infinity.

Next, the boundary condition at infinity is replaced by one at a finite distance $\bar{r}_{\text{max}}$, and a grid of $N + 1$ points $\bar{r}_i$ from the origin to $\bar{r}_{\text{max}}$ is used to construct a finite-difference approximation[15]

$$-\frac{u_{l,i+1} - 2u_{l,i} + u_{l,i-1}}{(\Delta r)^2} + \left[\frac{l(l+1)}{\bar{r}_i^2} + \zeta_N \bar{V}_{\text{eff}}(\bar{r}_i) \right] = \zeta_N \bar{E}_l u_{l,i}, \tag{8.45}$$

where $u_{l,i} = u_l(\bar{r}_i)$ and $\Delta r = \bar{r}_{\text{max}}/N$. Once again we have a matrix eigenvalue problem, with $u_{l,i}$ the components of a vector. In this case the matrix is tridiagonal. The last step is diagonalization of the matrix by standard means.[16]

8.4 *Exercises*

The exercises below give some guidance in the exploration of the physics that can be studied with HATOM. They include projects that require modification of the program. Before trying the exercises, review the next section, where the operation of the program is discussed.

8.1 **Hydrogen**
Verify that the numerical methods used for states distorted by electric and magnetic fields produce reasonably good results for undistorted hydrogenic ions. Select **Distorted Eigenfunctions** under **Compute** and, in the input screen that is presented, select manual mode and zero field strengths. The results obtained under these conditions can be compared with those obtained when **Undistorted Eigenfunctions** is selected, but only for the lowest energy state for a given magnetic quantum number m. Compare wave functions visually and eigenenergies numerically for $m = 0$, 1, and 2.

8.2 **Radial Probability Density**
Extract the location of the maximum in the radial probability density $r^2 R_{nl}^2(r)$ for the electron in hydrogen from the plot of this function for a series of n values. Do the computation with the **Undistorted Eigenfunctions** selection under **Compute**. The radial probability density appears in a small plot in the lower part of the screen. If it is obscured by the larger plot, select the hot key for **Contract** to bring it into view. If the location of the maximum cannot be read from the screen with satisfactory accuracy, try printing the screen and measuring the location on paper. Careful selection

of the range of the calculation can also help. With what power of n does the radius of the maximum increase (or decrease)? Use the same approach to analyze the dependence of the maximum's location on the nuclear charge, with n fixed at 1.

8.3 Angular Dependence of the Probability Density

Study the electron probability density of hydrogen as a function of angle. Specifically, for $l = 1$, with $m = 0$ and 1, and for $l = 2$, with $m = 0$, 1, and 2, determine the polar angles for minima and maxima in the probability density. Use the **Expand** hot key to obtain a large plot of the probability density, and select the contour plot option in the **ModView** dialog box; these should provide the optimum viewing conditions. Working from a printed copy of the screen may also be helpful.

8.4 Binding Energies and Nuclear Mass

Compute the ratios of the binding energies for the hydrogenic atoms H, D, and He$^+$. The mass of the nucleus can be adjusted in the main dialog box of the **Undistorted Eigenfunctions** menu selection under **Compute**. Are the expected values of the ratios obtained? Why or why not?

8.5 Binding Energies and Nuclear Charge

Compute values of the binding energy for the hydrogenic atoms H, He$^+$, and Li^{2+} and compare their dependence on nuclear charge to the theoretical prediction of proportionality to the square of the charge. Does the change in nuclear mass have any important consequences for this comparison? Both the charge and the mass of the nucleus can be changed in the main dialog box of the **Undistorted Eigenfunctions** selection.

8.6 Mesonium

Determine the binding energies of the lowest states of mesonium, a muon (mass: 0.106 GeV/c^2) bound to a proton. Consider at least the 1S, 2S, and 1P states. The choice of the light particle must be changed from the default choice of an electron; this is done with the dialog box obtained on selection of **Particle** from the main menu. The calculation itself is done in the **Undistorted Eigenfunctions** mode.

8.7 Positronium

Adjust the mass of the "nucleus" to be the mass of a positron, and thereby study positronium,[12] bound states of an electron and a positron. The positron is the antiparticle of the electron and therefore has the same mass as the electron (0.511 MeV/c^2), but the opposite charge. What are the masses of the lowest states? Consider at least the 1S, 2S, and 1P states. Keep in mind that the program computes only the binding energy.

8.8 Muonium

Repeat Exercise 8.7 for muonium, bound states of a muon and an electron. Use a positively charged muon as the nucleus; its mass is 0.106 GeV/c^2.

8.9 Linear Zeeman Effect

Compare the analytic value for the linear Zeeman energy shift (see Eq. 8.14), for $m = 1$, with that due to the quadratic term as computed by the program for a series of field strengths. Select **Distorted Eigenfunctions** under **Compute**; in the main dialog box that appears, select manual mode, zero electric field strength, and a non-zero magnetic field strength. For a weak field ($\ll 1\ MT$), either the basis-function method or the Lanczos method may be used.

8.10 Perturbative Zeeman Effect

Compare numerical results for the energy shift due to the quadratic Zeeman effect to first-order perturbation theory. (The quadratic Zeeman potential is included in Eq. 8.14.) At what field strength do they begin to differ by 10%? The use of the program to study the quadratic Zeeman effect is briefly described in Exercise 8.9.

8.11 Perturbative Stark Effect

Repeat Exercise 8.10 for the Stark effect in a uniform electric field. As was the case for the Zeeman potential, the Stark potential is contained in Eq. 8.14.

8.12 Variational Estimate for the Quadratic Zeeman Effect

Try simple variational wave functions such as $e^{-\alpha r}$ or $e^{-\beta r^2}$ for the state of the electron in the quadratic Zeeman problem, and compare the energies obtained with numerical results from the program.

8.13 Diamagnetic Susceptibility

Compute the diamagnetic susceptibility χ for hydrogenic ions by varying the B-field strength and comparing the field strength dependence of the energy to $-\frac{1}{2}\chi B^2$. This requires computation of the eigenenergy associated with the quadratic Zeeman term, for a series of field strength values, and a fit to a quadratic polynomial in B.

8.14 Polarizability

To leading order in perturbation theory, an electric field of strength $\mathcal{E}$ shifts the energy of a hydrogenic level by an amount $\frac{1}{2}\alpha\mathcal{E}^2$, where α is the polarizability. Extract an estimate for α in a way analogous to that used in Exercise 8.13. This requires computation of the eigenenergy associated with the Stark potential.

8.15 H_2^+ Binding

a. Determine the effective potential as a function of distance for two protons when an electron is in the lowest-energy state; this is done in the **Shared Eigenfunction** computation mode of the program. Select **Shared Eigenfunction** under **Compute**, and in the dialog box that appears, select auto mode (with the effective potential to be recorded), unit charges (to correspond to two protons), and a reasonable range of proton separation. The number of steps taken in auto mode must be set to provide acceptable resolution of the chosen range.

b. Use this potential to determine the binding energy, via the **Binding Of Molecule** mode. The **Binding Of Molecule** item in the **Compute** menu will become accessible once the effective potential has been computed. The nuclear masses chosen should be those of protons.

Experiments[3] done with H_2^+ find a binding energy of 2.8 eV; the minimum in the effective potential should be near 0.11 nm.

8.16 Variational Estimate for the H_2^+ Ion

Repeat Exercise 8.12 for the H_2^+ ion. The electron wave function must satisfy Eq. 8.16. A simple guess for a variational wave function in this case is [20,21] $e^{-\alpha r_1} + e^{-\alpha r_2}$, where $r_1 = |\mathbf{r} + \frac{1}{2}s\hat{z}|$ and $r_2 = |\mathbf{r} - \frac{1}{2}s\hat{z}|$ with s the nuclear separation. For some of the integrals, a change of variables to r_1, r_2, and the azimuthal angle can be useful. (See also the discussion by Park.[21]) The best values of the variational parameters, such as α in the example, will vary with s. Once the dependence on s is determined, the effective potential $V_{\text{eff}}(s)$, defined in Eq. 8.17, can be constructed and analyzed. At what separation is V_{eff} minimized? Compare this with the result obtained with the program for part a of Exercise 8.15.

8.17 Muomolecules

Consider the binding of H_2^+ when the electron is replaced by a muon.[22] Compare the binding energy and size of the muonic ion to the electronic one. For some discussion of the necessary steps, see Exercise 8.15. The replacement of the electron by the muon in the program is accomplished within the dialog box obtained by the selection of **Particle** from the main menu.

8.18 HHe$^+$ Binding

Can a single electron bind a proton to an alpha particle? Does a muon do a better job? Most of the steps necessary to answer these questions are briefly described in Exercise 8.15. The adjustments to be made are that the charge of the alpha particle is $2e$, not e, and the mass is roughly four times the proton mass. The charge is adjusted in the main dialog box presented when **Shared Eigenfunction** is selected, and the mass is entered in the dialog box associated with the **Binding Of Molecule** step. Also, the correct light particle, electron or muon, is selected from the **Particle** dialog box. Of course, the existence of binding is determined by the sign of the energy eigenvalue, and its strength by the magnitude. If, however, the size of the computational mesh is chosen to be too small, the eigenenergy will be artificially elevated. The visual test for a mesh that is much too small is the lack of an exponential tail in the wave function; the edge of the mesh will instead bring the wave function abruptly to zero. A complete calculation should include a study of convergence as the mesh size is increased.

8.19 Odd and Even Eigenfunctions

If the potential energy V is invariant under reflections, that is, if $V(-\mathbf{r}) = V(\mathbf{r})$, then the energy eigenfunctions ψ of the system can be classified as either odd or even under reflection: $\psi(-\mathbf{r}) = \pm\psi(\mathbf{r})$. The ground state, which has magnetic quantum number m of zero, is even. The lowest state with $m = 1$ is odd. With these facts as background, use the program to determine whether or not an odd electron state can lead to the binding of two protons into an excited state of H_2^+. The general approach to take is discussed in Exercise 8.15.

8.20 Molecular Binding and Nuclear Mass

Compare the binding energies of H_2^+, D_2^+, and HD$^+$. Which has the largest and which the smallest? The basic steps to begin the necessary calculations are described in Exercise 8.15. The mass of the deuteron is approximately twice that of the proton, and is entered in the dialog box presented by the **Binding Of Molecule** selection.

8.21 Molecular Excitations

Compare the scales of vibrational and rotational excitations to estimates obtained analytically under the assumptions of a rigid rotation decoupled from approximately harmonic oscillations in the radial direction. The vibrational excitations are associated with radial motion in the spherical effective potential and, consequently, with the quantum number n_r; this quantum number can be chosen in the dialog box presented when **Binding Of Molecule** is selected under **Compute**. Rotational excitations are associated with the angular momentum quantum number l, which can be chosen in the same dialog box. For the analytic estimate, the energy operator is viewed as separating into two pieces,[23] a vibrational piece $p_r^2/2\mu_N + V(r_0) + \frac{1}{2}(r - r_0)^2(\partial^2 V/\partial r^2)_{r_0}$, where p_r is the radial momentum and r_0 is the equilibrium separation, and a rotational piece $L^2/2\mu_N r_0^2$. How many rotational levels exist between vibrational levels? What is the effect on the vibrational energies of an increase in the angular momentum of the rotational excitation?

Projects

8.22 Choice of Scales

Modify the program to automatically update the length and energy scales L_0 and V_0 to accommodate changes in the nuclear charge Z, the principal quantum number n, and the mass of the light particle. Verify that plots of excited hydrogenic states will then automatically scale to the appropriate range.

8.23 Estimation of the Dipole Moment

Change the program to use the basis-function formula in Eq. 8.43 to estimate the dipole moment. Compare results with those obtained by the program from the finite-difference formula in Eq. 8.42.

8.24 Expectation Values

Add the computation of expectation values other than the dipole moment; use the segments of code for the dipole moment as guides. For example, compute the expectation values of the radial and angular positions of an electron in hydrogen distorted by external electric and magnetic fields.

8.25 Parity Eigenstates

When the nuclear charges of a diatomic molecule are equal, the electron wave function will be an eigenfunction of reflection, or *parity*. Extend the program to include computation of the lowest energy for both odd and even parity. To do this, restrict the computational grid to non-negative z and require the wave function in the excluded region to be determined by the chosen parity. Investigate the likelihood of binding by an electron in an odd parity state.

8.26 Distortion by a Coulomb Field

Modify the sections of the program that deal with hydrogenic wave functions distorted by electric and magnetic fields to include a Coulomb field generated by a charge placed on the z-axis but outside the computational grid. Study the distortion due to this external charge as a function of its distance, magnitude, and sign. Consider, in particular, the energy and dipole moment of the electron state. For these calculations, the finite-difference/Lanczos algorithm is readily adapted. One could also consider the basis-function approach if numerical quadrature is used for the necessary integrals; try Gauss-Laguerre integration.[24]

8.27   Charmonium

Make the necessary changes to the program to permit study of the eigenstates of charmonium, bound states of a charm quark and its antiparticle. Replace the Coulomb potential with the Cornell model[25] for the heavy quark-antiquark potential $V(r) = -g/r + a^2 r$. The parameters are $g = 0.52\hbar c$ and $a = 0.427\, \text{GeV}/\sqrt{\hbar c}$, and the charm quark mass is $m_c = 1.84\, \text{GeV}/c^2$. Use the matrix diagonalization method to solve the radial Schrödinger equation; this is the same method already used in the program to compute the binding wave function. Compare the computed masses to spin averages of the measured masses[26]: 1S, 3,068 MeV/c^2; 2S, 3,663 MeV/c^2; 1P, 3,525 MeV/c^2.

8.28   Heavy Baryons

A heavy baryon consists of three quarks, of which one or more is heavy. If two are heavy and the third relatively light, one can consider a Born–Oppenheimer approximation to the three-body bound state. To be specific, consider a bbs system: two bottom quarks and a strange quark. The strange quark is much lighter than the other two, but not so light that the nonrelativistic Schrödinger equation cannot be used. The two bottom quarks are then bound together by the strange quark and their direct interaction.

Estimate the mass of such a baryon by first using HATOM to compute the binding energy. This computation requires two steps, one to compute the effective potential between the two bottom quarks and another to solve the Schrödinger equation for the effective potential. Both steps require modification of the program to replace the ordinary Coulomb interaction with an appropriate model for the quark interaction. The Cornell model (see Exercise 8.27) is a reasonable choice. The quark masses can be taken to be 500 MeV/c^2 for the strange quark and 5 GeV/c^2 for the bottom quark.

8.29   Two-Dimensional Problems

The program can be modified to solve the Schrödinger equation in two Cartesian dimensions

$$-\frac{\hbar^2}{2\mu}\left(\frac{\partial^2}{\partial x^2} + \frac{\partial^2}{\partial y^2}\right)\psi + V(x, y)\psi = E\psi.$$

Two simple potentials are the square well, perhaps combined with a perturbation proportional to xy, and the anisotropic oscillator, with an optional anharmonic term such as $xy(x^2 + y^2)$. Carry out the necessary modifications for one of these potentials and obtain the ground state energy and wave function.

8.30   Two-Body Problems

A two-body problem in one dimension has the same number of degrees of freedom as a one-body problem in two dimensions, and the Schrödinger equations take nearly the same form in the two cases. The program can therefore be modified to treat the two-body problem.[*] An interesting interaction is that of two oscillators coupled by a linear force. Another is that of two "atoms," modeled as two particles bound to two adjustable centers of attraction. Notice that if two coupled oscillators have the same frequency, the two-dimensional equation can be separated into an analytically soluble pair of one-dimensional equations; this can provide a useful check on the results of the modified program. One can consider distinguishable particles, or fermions or bosons. For identical particles, the interactions must be invariant under interchange of their coordinates x_1 and x_2; calculations may then be restricted to $x_1 \geq x_2$. Select one of these cases and modify the program accordingly. Use the new program to compute energies and wave functions. In the case of two atoms, investigate the possibility of binding by computing the effective potential between the centers of attraction as a function of their separation.

8.5 *Details of the Program*

8.5.1 **Running the Program**

The program is controlled by choosing menu options. These options are as follows:

- **File:** Get program information, read and write data files, exit the program.
    - **About CUPS:** Show description of software consortium.
    - **About Program:** Show credits and a brief description.
    - **Configuration:** Verify and/or change program configuration.
    - **New:** Set file name to default file name and start new calculation.
    - **Open...:** Open file and read contents.
    - **Save:** Save current state of the program to a file.
    - **Save As..:** Save current state of the program to a file with chosen name and set file name to this choice.
    - **Play Back...:** Play recorded sequence.
    - **Exit Program**

- **Particle:** Select light particle type and mass, and units for energy and length.

- **Compute:** Plot previous results, if any, and compute new results, if desired, in the chosen category.
    - **Undistorted Eigenfunctions**
    - **Distorted Eigenfunctions:** External fields are present.
    - **Shared Eigenfunction:** Eigenfunction of shared light particle that provides molecular binding. An effective nuclear potential can be computed for a sequence of separations.
    - **Binding Of Molecule:** Compute the binding energy and molecular wave function.

---

[*]The time evolution of two-body systems can be observed with the quantum mechanics program IDENT.[27]

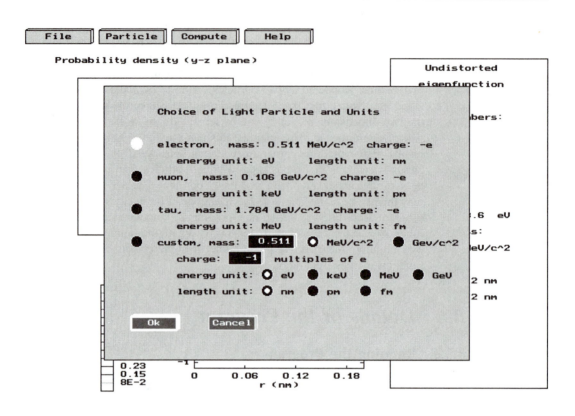

Figure 8.1: Input dialog box for the selection of the light particle and of the energy and length units.

- **Help:** Display help screens.
  - **Summary:** Display summary of menu choices.
  - **File:** Describe entries under **File**.
  - **Particle:** Describe input of particle properties.
  - **Compute:** Describe entries under **Compute**.
  - **Algorithms:** Describe algorithms.

The options under **Compute** lead to input screens for the selection of relevant parameters. Also within these options, certain keys are activated for control of the display and of calculations. Some keys bring up input screens that request additional information, such as choice of a display type for the wave function.

8.5.2   Sample Input and Output

Figure 8.1 shows the dialog box used for the selection of the light particle and of the energy and length units used by the program. Typical output of the program is shown in Figures 8.2–8.5. The wave functions may be shown as contour plots or as surface plots.

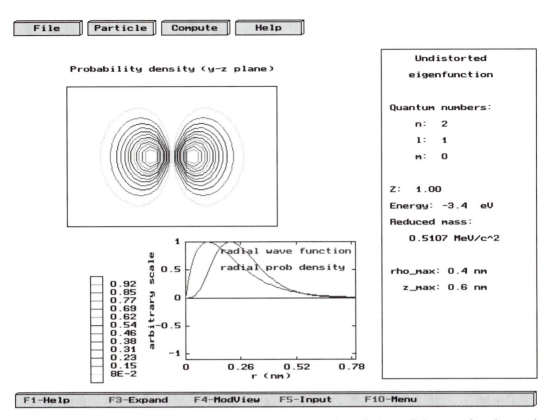

Figure 8.2: An excited state of hydrogen. The smaller plot of the radial wave function and probability density uses color for identification of the two functions.

References

1. Rajagopal, A.K., Chanmugam, G., O'Connell, R.F., Surmelian, G.L. "Ionization Energies of Hydrogen in Magnetic White Dwarfs," *Astrophysical Journal* **177**(3):713, 1972.

2. Rau, A.R.P., Spruch, L. "Energy levels of Hydrogen in Magnetic Fields of Arbitrary Strength," *Astrophysical Journal* **207**(2):671, 1976 and references given therein; Clark, C.W., Taylor, K.T. "The Quadratic Zeeman Effect in Hydrogen Rydberg Series: Application of Sturmian Functions," *Journal of Physics B: Atomic and Molecular Physics* **15**(8):1175, 1982.

3. Gasiorowicz, S. *Quantum Physics*. New York: John Wiley & Sons, 1974.

4. Bates, D.R., Ledsham, K., Stewart, A.L. "Wave Functions of the Hydrogen Molecular Ion," *Philosophical Transactions of the Royal Society of London*

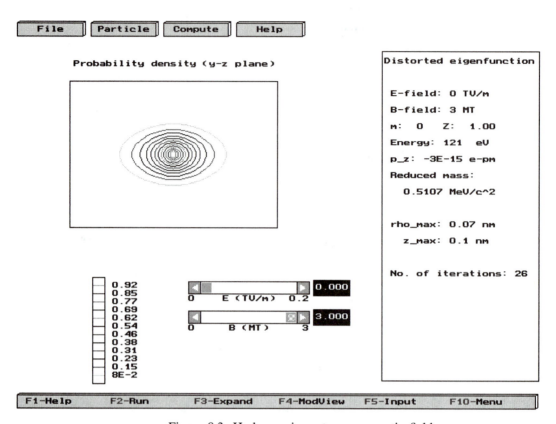

Figure 8.3: Hydrogen in a strong magnetic field.

A246(911):215, 1953; Liu, J.W. "Analytic Solutions to the Generalized Sphe-riodal Wave Equation and the Green's Function of One-Electron Diatomic Molecules," *Journal of Mathematical Physics* 33(12):4026, 1992.

5. Lanczos, C. "An Iteration Method for the Solution of the Eigenvalue Problem of Linear Differential and Integral Operators," *Journal of Research of the National Bureau of Standards* **45**:255, 1950; Cullum, J., Willoughby, R.A. "Computing Eigenvalues of Very Large Symmetric Matrices—An Implementation of a Lanczos Algorithm with no Reorthogonalization," *Journal of Computational Physics* **44**(2):329, 1981; Cullum, J., Willoughby, R.A. *Lanczos Algorithms for Large Symmetric Eigenvalue Computations*, Vols. I and II. Boston: Birkhauser, 1985; Scott, D.S. "The Lanczos Algorithm," in *Sparse Matrices and their Uses*, ed. I.S. Duff, pp. 139-159. London: Academic Press, 1981.

6. Anderson, E.E. *Modern Physics and Quantum Mechanics,* Philadelphia: Saunders, 1971.

7. Eisberg, R., Resnick, R. *Quantum Physics of Atoms, Molecules, Solids, Nuclei and Particles,* 2nd ed. New York: John Wiley & Sons, 1985.

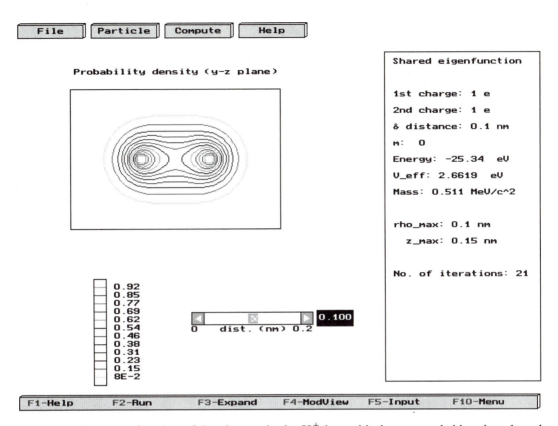

Figure 8.4: The wave function of the electron in the H_2^+ ion, with the protons held at the selected separation.

8. Park, D.A. *Introduction to the Quantum Theory,* 3rd ed. New York: McGraw-Hill, 1992.

9. Schmid, E.W., Spitz, G., Lösch, W. *Theoretical Physics on the Personal Computer.* Berlin: Springer-Verlag, 1988.

10. Merzbacher, E., *Quantum Mechanics,* 2nd ed. New York: John Wiley & Sons, 1970.

11. Palmer, W.F., Taylor, R.J. "Separability of Center of Mass and Relative Motion of Hydrogen in Very Strong Magnetic Fields," *American Journal of Physics* **49**(9):855, 1981.

12. Bethe, H.A., Salpeter, E.E. *Quantum Mechanics of One- and Two-Electron Atoms.* Berlin: Springer-Verlag, 1957.

13. Schiff, L.I., Snyder, H. "Theory of the Quadratic Zeeman Effect," *Physical Review* **55**(1):59, 1939.

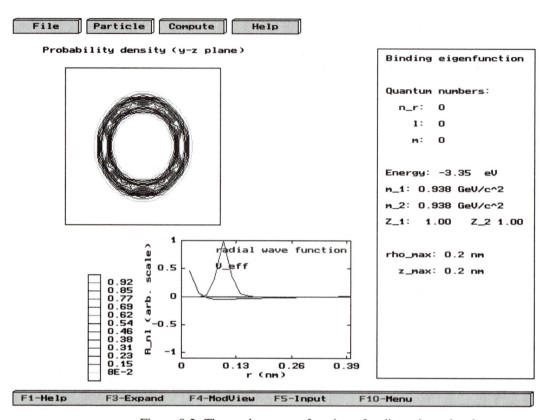

Figure 8.5: The nuclear wave function of a diatomic molecule.

14. Gasiorowicz, S. *Quantum Physics.* New York: John Wiley & Sons, 1974, pp. 314-315.

15. Gerald, C.F., Wheatley, P.O. *Applied Numerical Analysis,* 4th ed. Reading, MA: Addison-Wesley, 1989.

16. Press, W.H., Teukolsky, S.A., Vetterling, W.J., Flannery, B.P. *Numerical Recipes: The Art of Scientific Computation*, 2nd ed. Cambridge: Cambridge University Press, 1992.

17. Paige, C.C. "Practical Use of the Symmetric Lanczos Process with Re-orthogonalization," *BIT (Copenhagen)* **10**(2):183, 1970; "Computational Variants of the Lanczos Method for the Eigenproblem," *Journal of the Institute of Mathematics and Its Applications (London)* **10**(3):373, 1972; Parlett, B.N., Scott, D.S. "The Lanczos Algorithm with Selective Orthogonalization," *Mathematics of Computation* **33**(145):217, 1979; Parlett, B.N., Nour-Omid, B. "Towards a Black Box Lanczos Program," *Computer Physics Communications* **53**(1-3):169, 1989.

18. Scott, D.S. "Implementing Lanczos-Like Algorithms on Hypercube Architectures," *Computer Physics Communications* **53**(1-3):271, 1989.

19. Brandi, H.S. "Hydrogen Atoms in Strong Magnetic Fields," *Physical Review A* **11**(6):1835, 1975.

20. Davydov, A.S. *Quantum Mechanics.* Reading, MA: Addison-Wesley, 1965, p. 478.

21. Park, D.A. *Introduction to the Quantum Theory*, 3rd ed. New York: McGraw-Hill, 1992, p. 503.

22. Bracci, L., Fiorentini, G. "Mesic Molecules and Muon Catalyzed Fusion," *Physics Reports* **86**(4):169, 1982.

23. Liboff, R.L. *Introductory Quantum Mechanics,* 2nd ed. San Francisco: Holden-Day, 1992, p. 608.

24. Abramowitz, M., Stegun, I.A. eds. *Handbook of Mathematical Functions.* New York: Dover, 1964.

25. Eichten, E., Gottfried, K., Kinoshita, T., Lane, K.D., Yan, T.-M. "Charmonium: The Model," *Physical Review D* **17**(11):3090, 1978; "Charmonium: Comparison with Experiment," *ibid.,* **21**(1):203, 1980.

26. Particle Data Group, Hikasa, K. *et al.,* "Review of Particle Properties," *Physical Review D* **45**(11):Part II, June 1992.

27. Hiller, J.R., Johnston, I.D., Styer, D.F. *Quantum Mechanics Simulations.* New York: John Wiley & Sons, 1995.

Appendix

Walk-Throughs for All Programs

These "walk-throughs" are intended to give you a quick overview of each program. Please see the Introduction for one-paragraph descriptions for all programs.

A.1 Walk-Through for QUANTUM Program

- **Change the sliders beneath the graphs and view the changes in the probability densities.** The initial screen places the program in the **Uncertainty Principle** section of the program.

- **Press the F6-Show Real hot key to toggle on and off the real and imaginary components of the wave function in the display.** Change the mean position of the spatial wave function with the real and imaginary components displayed to observe how the mean position of the wave packet is related to the phase of the momentum space wavefunction.

- **From the Section menu, select Time-Dependent Free Particles.** This places the program in a mode to view the time evolution of a free wavepacket. Parameters of the wavepacket can be chosen with the appropriate slider shown on the screen.

- **Press the F2-Run/stop hot key.** The wavepacket time evolution will displayed in the graph shown. Again the F6 hot key will toggle between displaying and not displaying the real and imaginary components of the wavefunction. Press **F2-Run/Stop** to stop the animation.

- **From the Section menu, select the Barriers item.** This selection places the program in a mode to study solutions to problems when the desired solutions are asymptotically free. A potential barrier is displayed in the graph along with a Hamiltonian eigenwave.

- **Change the energy of the eigenwave using the energy slider.** This changes the value of the energy and a new eigenwave is calculated and displayed. The reflection and transmission graph shows a vertical gray line that represents the energy corresponding to the currently selected eigenwave.

- **Select Barrier Parameters from the Parameters menu.** This selection brings up an input screen that allows the user to input the number of constant potential regions and the potential in each region.

- **Change the barrier as desired or press Cancel to cancel this input screen.**

- **Select Particle Parameters from the Parameters menu.** This allows the user to select parameters of the eigenwave or to select a wave packet to display.

- **Select Wave Packet and accept the input screen.** The program will calculate a wavepacket from component eigenwaves and display it on the screen. Also, the reflection and transmission graph shows the distribution of energy eigenvalues included in the wave packet with a gray curve.

- **Press the F2-Run/Stop hot key to begin animating the wave packet.**

- **Press the F2-Run/Stop hot key again to stop the animation.**

- **Select Bound Solutions from the Section menu.** This places the program in a mode to investigate bound solutions to Schrodinger's equation. Initially a square well is shown in the left graph along with a wavepacket constructed of up to three eigensolutions. The right graph shows the eigenwaves for up to the lowest ten Hamiltonian eigenvalues. The table at the far right shows the eigenvalues.

- **Press the F7-Integral hot key.** This brings up an integration tool that allows the user to find several different integrals over selectable limits. Select probability density and limits 0 to 1 and press **OK** to see the integral.

- **Select Cancel to cancel the integration tool display.**

- **Select Well Parameters from the Parameters menu.**

- **Select User defined potential and accept the default text string for the potential.** In this mode the program searches for the eigensolutions to the Hamiltonian.

- **Press "s" or "S" to begin the search for eigenvalues.** The program displays the trial eigenwaves as it searches for eigenvalues. "Q" or "q" may be pressed to suspend or abandon the search.

A.2   Walk-Through for SPECREL Program

The initial screen show two graphs and the world lines of a stick and two flashes.

- **Click the Run/Stop button in the center of the screen or press the F2 key.** The current graphs shown are x versus y versus t graphs. Clicking the run button causes the program to increment the time parameter and the graphs show the defined objects location on a constant t plane.

- **Click the Run/Stop button or press the F2 key again to stop animation** This stops the animation and the graphs can be changed, the object list edited, or the motion of frame 2 changed.

- **Click on the buttons labeled xy, xt, xyt, and xyz along the upper left edge of the display.** These buttons change the left graph to display the coordinates selected. The buttons along the upper right edge of the screen change the coordinates of the right graph. The graph type can also be selected from the menu Graph 1 or Graph 2 if the user does not have a mouse.

- **Click on the buttons labeled 1 and 2 near the bottom edge of the graph at the left edge of the display.** These buttons select which reference frame is displayed on the left graph. The corresponding buttons on the right edge of the display select which reference frame is displayed in the right graph. The graph type can also be selected from the menu **Graph 1** or **Graph 2** if the user does not have a mouse.

- **Click or click and hold on the up button below the window labeled β_x. Watch the values length and velocity of object 1 shown on the index card in the lower left of the screen change as β_x changes.** This changes the motion of reference frame 2 relative to reference frame 1. When the button is released, the program delays about 1 second to check if you make further frame changes; if no more changes are made it recalculates the trajectories of the objects in reference frame 2 and updates the appropriate graph.

- **Click on the Next button in the column of buttons along the lower left edge of the screen.** This brings the information card for the next defined object to the front so that its information can be viewed and edited if desired.

- **Click on one of the fields of information shown on the displayed information card, press the Edit button, or press the F3 key.** This places the program in edit mode, and the value in the currently highlighted field can be changed using the number keys. **Dimension changes** change the dimension values for all reference frames. Velocity changes in one frame change the velocity in the other frame. The highlighted field can be changed using the **Tab** key, clicking on the **Next** button, or clicking on the desired field.

- **Click on the Edit button, press F3, or press Enter to end edit mode.** This will end edit mode and new trajectories will be calculated and the graphs updated.

- **Select Four Vectors from the Objects menu.** After selecting this the program displays the components of a four vector on the information card and displays a graph of the four vector in the graphs. The program is always able to edit the four vector components and one component is always highlighted while four vectors are displayed. The highlighted component can be changed by pressing the **Tab** key or clicking on the desired component.

- **Click or click and hold the Up or Down button on the side of the information card on the lower left edge of the display.** This increments or decrements the value of the highlighted component by 0.01, respectively. Changes to the affected components in the other reference frame are made automatically and the two graphs are updated.

- **Select 2nd Rank Four Tensors from the Objects menu.** After selecting this item, the components of a 2nd rank four tensor are displayed for both reference frames in the information card. These behave in a manner similar to the four vectors. The symmetry of the tensors can be forced by clicking on the appropriate checkbox at the bottom of the information card. Antisymmetric tensors are represented in the display as two space vectors in analogy to electric and magnetic fields. This makes the program useful for seeing how electric and magnetic fields transform. Nonsymmetric and symmetric tensors are displayed as the distortion of a cube. Currently, the display of those types of tensors has some errors in it.

A.3 Walk-Through for RUTHERFD Program

- **Press-F2 Run/Stop to begin generating results of scattering events calculated from random impact parameters.** The program begins with a inverse square scattering force. Collect enough events to be able to see the profile of the scattering cross section on the display. The **Detail Histogram** can be used to look at the data generated in more detail.

- **Select regular NGon from the Force menu.** This item brings up an input screen to select the number of sides of the regular hard regular polygon scatterer and its diameter.

- **Input 5 for the number of sides and a diameter of 1 and select OK. Press F3 and OK to clear data previously collected.** Note: This will be changed to automatically clear previous data when a new force is selected.

- **Press F2-Run/Stop to collect data. Press F2-Run/Stop again to stop collecting data. Change the orientation slider to a new direction.** This changes the orientation of the force center to assist in attempting to determine the symmetry of the scattering force. Press F2 again to collect some more data and F2 again to stop collecting data.

- **Select Guess the Scatterer from the Mode menu.** This puts the program into a mode in which the user does not know the scattering force, but can do scattering simulations to try to determine the force. Try collecting some quickly to see if it tells you something about the scattering force.

- **Press the F7 key to reveal what the scattering force is.** This shows the form of the scattering force and any parameters needed to fully define the force. Note: Additional help screens are being created to help the user get to a solution to the unknown scatterer. Press the F8 key to get a new unknown scatterer.

- **Select Geiger-Marsden Experiment from the Mode menu.** This initially displays a diagram of the Geiger-Marsden experiment performed to discover Rutherford scattering.

- **Click the mouse or press any key.** This program displays a scintillation screen and some sliders. The user can get some feel for what was involved in collecting data by this method.

A.4   Walk-Through for HATOM Program

The initial screen comes up with a plot of the wave function for the ground state of hydrogen.

- **Select Undistorted Eigenfunctions from the Compute menu.** An input screen appears. Enter new quantum numbers such as $(2, 1, 1)$ for (n, l, m), and increase $r_{\max}$ to 0.6 nm.

- **Press F3 to expand the contour plot to full-screen size.** Use F4 to obtain a dialog box from which alternate views may be chosen. In the case of a surface plot, checking **Modify viewpoint?** brings a special input mode for the adjustment of the view of the surface. The three Euler angles are varied using sliders, and the size and distance of the plot are changed with hot keys F3 and F4. Any changes are immediately reflected in a small box in the lower left, and are applied to the wave function plot when F2 is pressed. The default viewpoint can be recovered by pressing F5. When finished adjusting the viewpoint, press **Enter**.

- **Select Shared Eigenfunctions from the Compute menu.** In the dialog box, select **auto** as the mode, check **record to file** and **record effective potential**. Set the magnetic quantum number to 0. Set both ρ_{max} and z_{max} to 0.2 nm.

- **Accept the file name of HATOM.DAT.** If, however, that would mean losing something saved previously, change the name.

- **Observe the series of iterations.** The program will automatically step through a sequence of nuclear separations and at each value will compute the electron wave function and the value of the effective potential. The computation of each wave function is done iteratively, and the iteration count and latest energy estimate are displayed near the middle of the screen.

- **Select Play Back . . . from the File menu.** When the dialog box appears, press **Enter** to continue.

- **Select Binding Of Molecule from the Compute menu.** Enter quantum numbers of $(0, 0, 0)$ for (n_r, l, m). The program then calculates and plots the ground-state wave function associated with the relative motion of the nuclei. The calculation is somewhat crude, because so few points are used; however, the result of 3.25 eV for the binding energy is not far from the experimental result of 2.8 eV. Also, the minimum in the potential is at ~ 0.1 nm, which is to be compared with an expected value of 0.11 nm.

- **Select Exit Program from the File menu.** The program will not allow an immediate exit if calculations have been done but not saved. There is instead an opportunity to save the work to a file. In this case, saving the work will mean saving the molecular binding calculation. One might instead wish to retain the sequence of shared eigenfunctions, which takes much longer to calculate and which will be overwritten if the last calculation is saved using the same file name. Saving calculations can be done directly with other entries in the **File** menu. Notice that work saved previously, perhaps in preparation for a lecture demonstration, can be recalled with the **Open . . .** or **Play Back** entries.

A.5 Walk-Through for GERMER Program

Under **Runs**, select **Davisson-Germer Scattering**. This screen displays scattering of an electron beam to a detector. The top two sliders control accelerating voltage; the middle two control limits of the detector travel. The bottom slider adjusts target crystal orientation. Press **Tab** twice. This moves the slider control to the top slider, the one which lets you change the electron accelerating voltage.

- **Angle Scan.** Using just two keys (more convenient than the mouse in this case), you can quickly simulate the classic 1927 Davisson-Germer experiment. Press the right arrow key several times to get the voltage to, say, 48 V. Then press F5, and you will see the detector sweep through its angular range, showing some small structure at around 130°. Press the right arrow key a twice, followed by F5 to do a run at about 50 V. Succeeding runs at 2-V intervals will let you see how the structure at 130° turns into a major peak when the accelerating voltage is in the vicinity of 54 V; then the peak recedes at higher accelerating voltages. [The major peak in the original 1927 data was at 50° clockwise from the incident direction (a "scan angle" of about 130°)].

- **Voltage Scan.** This simulates some experiments done with the detector at a fixed angle while the voltage is varied (the square root of voltage is varied by a factor of four in this mode).

Walk-Through G. P. Thomson Scattering

Under **Runs**, select **G. P. Thomson Scattering**. This screen displays scattering of an electron beam to a detector, though in fact G. P Thomson's "detector" was photographic film. Notice that the range of accelerating voltage is much higher than in the Davisson-Germer experiments if the electron beam is to penetrate the thin metal sheet and reach the film plane. The range of detector travel is also smaller, with major interest focused on the near-forward direction. The bottom three sliders control the "metal film" parameters. The crystallites of the "film" are randomly oriented within 0.05° of the selected "center" angle; their number can be varied, as well as the number, N, of atoms in each row or column of each $N \times N$ crystallite. The **Planes** hot key toggles between crystals whose scattering planes are ($hk0$) and those whose scattering planes are (111).

- **Angle Scan.** The default setup is for a single crystal with a fairly low-energy electron beam coming in. Select **AngScan** and you will see the detector move through an angular range, and detect two major scattering peaks.

- **Number in a Row or Column (N).** The default crystal has six atoms in each row or column (6 × 6). Change the slider for N to its minimum value of 2, and do another angular scan. Then change N to 25 or so and do another angular scan. This will give you some feeling for the effect of crystal size on the sharpness of scattering peaks.

- **Number of Crystallites.** Change this slider to 10, and observe that 10 different crystallites are drawn. Now do an angular scan. The results of angular scans are sensitive to both the number of crystallites, and the number of N in each row or column. The rings in G. P. Thomson's pictures were due to scattering from groups of crystallites oriented at or near the correct angle for Bragg scattering from a given crystal plane.

The X-Ray Diffractometer.

Under **Runs**, select **X-Ray Diffractometer**. This screen simulates the operation of an x-ray diffractometer, where the scattering crystal is always oriented so that Bragg-scattered photons will reach the detector. For diffractometers with copper targets, the x-ray spectrum is dominated by the K_α and K_β lines of copper (8.04 keV and 8.90 keV, respectively). The walk-through will let you see these two dominant peaks in a typical diffractometer.

- **The K_α peak.** Select a total accelerating voltage of around 8,000 V, and do an angle scan by clicking the hot key or pressing F5. You will see the largest peak in the spectrum, that due to the K_α x-ray when a vacancy in the K-shell is filled by an electron falling from the L-shell. This occurs near a detector angle of 45° (a Bragg scattering angle of 22.5°).

- **The K_β peak.** Move the total accelerating voltage to around 9,000 V, and redo the angle scan. The peak shifts to lower angle (around 40°) at higher energies and shorter wavelengths. In this case the K-shell vacancy is being filled by an M-shell electron.

A.6   Walk-Through for LASER Program

- **Populations.** Populations of the three levels are displayed in yellow at the right-hand edge of the screen. The nominal total population is 400 atoms, and initially most atoms are in the ground state.

- **Flow Rates.** Transitions per second between levels are shown by the six vertical bars representing spontaneous and stimulated transitions between the three pairs of levels. Spontaneous transitions are shown on the **Left** and stimulated transitions on the **Right** for each pair. Numerical values are given below the bars.

Under **Laser**, select **Flow Rates: Lasing 2→1**. This screen displays flow rates between the three levels of the laser as one changes the laser parameters.

Make the A_{21} Einstein coefficient smaller by dragging it to the left until its log is something like -9. This makes the lifetime longer for spontaneous decay from level 2 to level 1, so atoms will accumulate in level 2. Note that the level populations don't immediately change, since at thermal equilibrium populations depend on level spacing only.

Increase the pumping power (bottom slider) until green transition arrows are visible (log power about 6) between levels 2 and 0, showing that pumping is being equalled by spontaneous decay. Gradually increase power until lasing begins, as shown by a red arrow (log power around 7.5–8). Now there is a net flow rate of atoms into level 2 from level 0, balanced by the stimulated emission from level 2 to level 1.

Try adjusting the mirror transmission, and observe the decrease of lasing intensity within the cavity due to some radiation leaving through the mirror. To see a graph of output vs. mirror transmission, select **Do Plot** and choose **Output Intensity versus Mirror Transmission.** All settings remain in place when you do this graph, so you may immediately return via **Lasers** and **Flow Rates** to continue exploring the effects of changing parameters on laser behavior.

Walk-Through Cavity Stability
Select **Cavity**, then **Cavity Stability Via Ray Tracing**. Insert a ray into the cavity by pressing F2 or activating the F2 hot key at the bottom of the screen. For the initial cavity configuration, the rays remain near the axis, showing the cavity to be a stable one. The shape of the gaussian beam within the cavity is suggested by the envelope of the rays. Note that the color scale at the top of the screen shows the number of passes made by the rays over a point on the screen. When a point on the screen is white, 15 or more passes have been made over that spot.

Go to **Cavity** and **Set Cavity Parameters** to adjust the mirror radii, and distance between mirrors. Decrease the radius at either end until the cavity becomes unstable.

A.7   Walk-Through for NUCLEAR Program

A.7.1   Properties of Nuclei

Walk-Through Mass Formula
Under **Pick Plot**, select **Mass Formula: N or A vs. Z**
Select **Do Plot** from the bar menu and observe how well a particular semi-empirical mass formula (SEMF) fits the existing data on all nuclides. The blue areas are where the nuclides are more stable (more binding energy/nucleon) than the SEMF predicts. The green areas are where the SEMF and binding energy/nucleon agree to within 0.01 MeV/nucleon. The red areas are where the SEMF predicts more (by at least 0.01 MeV/nucleon) binding energy/nucleon than the nuclei possess.

The two horizontal blue lines suggest special stability in the vicinity of 50 and 80 neutrons. The largest blue region is roughly at the junction of 80 protons, and 125 neutrons, and the next largest blue region is where roughly 80 neutrons intersect 50 protons. This is a rather high-level view of the "magic numbers" 50, 82, and 126.

You may wish to return to the SEMF bar menu and select only stable nuclei for a plot, or you may wish to plot A versus Z, and you will probably be interested in changing the "resolution" of 0.01 MeV/nucleon.

Walk-Through Decays and Reactions
Under **Pick Plot**, select **Nuclear Reactions: N or A vs. Z**.

- **Alpha Decays.** Select **Decay Type** and a fresh bar menu will appear. Select **Emit α**. After this, put in a **Minimum Particle Energy** for the emitted alphas of 4 MeV, then **Plot the Graph**. The graph shows where the nuclei are located which are energetically capable of emitting alpha particles with an energy of at least 4 MeV. There are (very light colored) square cursors at each end of the data in the plot. You can use the mouse to drag a cursor, or arrow keys to move cursors, and then **Zoom In** on the area between the cursors for a better view.

- **Neutron Energies.** This time for **Decay Type** select **Emit n**, and put in a **Minimum Particle Energy** of -5 MeV. This strange-looking value lets you see how tightly bound the last neutron is in a given nucleus. It will display all neutrons which are bound by 5 MeV or less. The plot in fact shows several regions, all fairly neutron-rich, which have horizontal, finger-like regions showing. Zoom in on one or two of these regions and note that it is the odd-N nuclei which have a final-neutron binding energy less than 5 MeV. This suggests that odd numbers of neutrons are associated with a decrease in binding energy (the average for the whole nucleus is well over 7 MeV/nucleon, after all).

- **Proton Energies.** After looking at the effect of odd neutrons, what about the effect of odd protons? Is there one? Where will it be located? What would you do to find out? Will the binding energy you look for be the same, or more or less than the 5 MeV for neutrons?

Walk-Through Mass-Binding Energy Plots
Under **Pick Plot**, select **Mass or BE vs. N, Z, or A**.

Using the defaults and selecting **Plot the Graph** under this option gives you the "standard" view of the binding energy per nucleon versus the atomic number, Z. Stable nuclei are shown in light red, and other nuclei are in yellow. If you press PgDn, the cursor increment is changed to 100, and then if you press the **right arrow** key the left cursor moves up through 100 data points on the graph. Now if you **Zoom In**, the region between the cursors is replotted in the available space on the graph. As the cursor moves to new data points, the upper right-hand corner of the graph updates information about the nuclide where the "active" cursor is located.

If instead of selecting **Plot the Graph**, you select **at constant Z, N, A, nothing**, the item rotates from "nothing" held constant, to a constant number of neutrons (N) equal to 123. Now when you **Plot the Graph**, a rather parabolic set of data points appears, giving the binding energy per nucleon for all nuclides where $N = 123$. By pressing "P", you enter "parabola" mode and a parabola appears on the screen. You may move the three parabola cursors around on the screen by dragging with a mouse, or by using keystrokes. The equation of the parabola is available, and will make sense if the parabola is a close fit to the displayed graph.

A.7.2 Counting, Activation, and Decay Rates

This part of the program has two sections.

- **Section 1—Geiger Tube Counting.** Start a counting "run" in this section, and you will see a series of vertical bars appearing and being updated on the screen. These represent count distributions which would be obtained by (in blue) an ideal Geiger-Muller Tube (GMT) and (in cyan) by a GMT with dead time. You may interrupt the counting process or let it run to its conclusion. By using the **Poisson** hot key, you put a Poisson distribution on the screen along with the data, for you to move around on the screen with mouse or keys. The idea is to verify that the Poisson distribution is a very good fit to the "ideal" GMT data, but not so good for the "real" GMT data with resolving (dead) time effects present. The higher the count rate, the more the "real" GMT counts exceed the Poisson curve in the middle, and fall short on the edges. You may easily check this by going to **Set Parameters** on the top menu, and increasing the count rate to, say, 1,000 counts/s. This will also demonstrate a significant drop in the count rate itself, as the increased dead time takes its toll on the "live" counts detected.

- **Section 2—Decay and Growth Rates.** Under **Run** on the top menu in this section, select **Activation Run**. This plots the populations of radioactive nuclei in a material being "activated" in a neutron source (perhaps a neutron "howitzer"). The default is for silver, which has two stable isotopes. As neutrons are absorbed by the stable nuclei, radioactive nuclei are formed, and the populations of both radioactive species are plotted, along with the total population of radioactive nuclei. If the flux is left on during the whole run, the populations grow from zero and begin to level off. After a long time, all populations would become constant as the production by neutron absorption would be equalled by decay of the radioactive nuclei.

  Now do another activation run, but turn the neutron flux off in the middle of the run somewhere. When a run concludes, there is a **Pick Plot** item which appears on the top menu. Underneath this item, select **Ln(decay rate) vs. time**. The plot which appears shows the *ln* of total decay rate (which one would measure experimentally) and the decay rates from each species. After the flux is turned off, the individual decay curves appear to be nearly straight. One may use the cursors (on this graph, the cursors are vertical lines) to **Zoom In** on the region of interest, and then go into **Line** mode by pressing "L" so that a line will appear on the screen. The cursors at the end of this line can be moved around via the mouse or keystrokes, and the equation obtained by pressing End. This lets the user fit the line to a particular sections of data and to determine its slope.

  Now try a **Series Run** from the top menu (under **Runs**). This plots the series decay of species 1 into species 2, which then decays into species 3. The defaults are set up with species 3 having a long half-life compared to the plotting time, so its population grows throughout the plot. The population of species 2 is set up to go through a maximum, which may surprise some students, and lead to interesting investigations of production and decay rates.

  It is interesting to again select **Ln(decay rate) vs. time** under **Pick Plot** because it may be surprising that two species appear to have the same decay rates, while the third seems to be approaching a constant decay rate. One may also note that the log scale is needed in order to get an idea of the behavior of the smallest decay rate; on the plot of decay rates versus time, this lowest rate is right along the bottom of the graph.

Index

Limited Use License Agreement

This is the John Wiley & Sons, Inc. (Wiley) limited use License Agreement, which governs your use of any Wiley proprietary software products (Licensed Program) and User Manual(s) delivered with it.

Your use of the Licensed Program indicates your acceptance of the terms and conditions of this Agreement. If you do not accept or agree with them, you must return the Licensed Program unused within 30 days of receipt or, if purchased, within 30 days, as evidenced by a copy of your receipt, in which case, the purchase price will be fully refunded.

License: Wiley hereby grants you, and you accept, a non-exclusive and non-transferrable license, to use the Licensed Program and User Manual(s) on the following terms and conditions:

a. The Licensed Program and User Manual(s) are for your personal use only.
b. You may use the Licensed Program on a single computer, or on its temporary replacement, or on a subsequent computer only.
c. You may modify the Licensed Program for your use only, but any such modifications void all warranties expressed or implied. In all respects, the modified programs will continue to be subject to the terms and conditions of this Agreement.
d. A backup copy or copies may be made only as provided by the User Manual(s), but all such backup copies are subject to the terms and conditions of this Agreement.
e. You may not use the Licensed Program on more than one computer system, make or distribute unauthorized copies of the Licensed Program or User Manual(s), create by decompilation or otherwise the source code of the Licensed Program or use, copy, modify, or transfer the Licensed Program, in whole or in part, or User Manual(s), except as expressly permitted by this Agreement.
If you transfer possession of any copy or modification of the Licensed Program to any third party, your license is automatically terminated. Such termination shall be in addition to and not in lieu of any equitable, civil, or other remedies available to Wiley.

Term: This License Agreement is effective until terminated. You may terminate it at any time by destroying the Licensed Program and User Manual together with all copies made (with or without authorization).
This Agreement will also terminate upon the conditions discussed elsewhere in this Agreement, or if you fail to comply with any term or condition of this Agreement. Upon such termination, you agree to destroy the Licensed Program, User Manual(s), and any copies made (with or without authorization) of either.

Wiley's Rights: You acknowledge that the Licensed Program and User Manual(s) are the sole and exclusive property of Wiley. By accepting this Agreement, you do not become the owner of the Licensed Program or User Manual(s), but you do have the right to use them in accordance with the provisions of this Agreement. You agree to protect the Licensed Program and User Manual(s) from unauthorized use, reproduction or distribution.

Warranty: To the original licensee only, Wiley warrants that the diskettes on which the Licensed Program is furnished are free from defects in the materials and workmanship under normal use for a period of ninety (90) days from the date of purchase or receipt as evidenced by a copy of your receipt. If during the ninety day period, a defect in any diskette occurs, you may return it. Wiley will replace the defective diskette(s) without charge to you. Your sole and exclusive remedy in the event of a defect is expressly limited to replacement of the defective diskette(s) at no additional charge. This warranty does not apply to damage or defects due to improper use or negligence.
This limited warranty is in lieu of all other warranties, expressed or implied, including, without limitation, any warranties of merchantability or fitness for a particular purpose.
Except as specified above, the Licensed Program and User Manual(s) are furnished by Wiley on an "as is" basis and without warranty as to the performance or results you may obtain by using the Licensed Program and User Manual(s). The entire risk as to the results or performance, and the cost of all necessary servicing, repair, or correction of the Licensed Program and User Manual(s) is assumed by you.
In no event will Wiley be liable to you for any damages, including lost profits, lost savings, or other incidental or consequential damages arising out of the use or inability to use the Licensed Program or User Manual(s), even if Wiley or an authorized Wiley dealer has been advised of the possibility of such damages.

General: This Limited Warranty gives you specific legal rights. You may have others by operation of law which varies from state to state. If any of the provisions of this Agreement are invalid under any applicable statute or rule of law, they are to that extent deemed omitted.
This Agreement represents the entire agreement between us and supercedes any proposals or prior Agreements, oral or written, and any other communication between us relating to the subject matter of this Agreement.
This Agreement will be governed and construed as if wholly entered into and performed within the State of New York.
You acknowledge that you have read this Agreement, and agree to be bound by its terms and conditions.